Wind Tunnels: Models, Aerodynamics and Applications

Wind Tunnels: Models, Aerodynamics and Applications

Edited by **Russell Mikel**

CLANRYE
INTERNATIONAL

New Jersey

Published by Clanrye International,
55 Van Reypen Street,
Jersey City, NJ 07306, USA
www.clanryeinternational.com

Wind Tunnels: Models, Aerodynamics and Applications
Edited by Russell Mikel

International Standard Book Number: 978-1-63240-523-4 (Hardback)

Contents

Preface

The models, aerodynamics and applications of wind tunnels are discussed in this profound book. It will be a valuable tool for students and professionals. It provides an insight into various designs of wind tunnelling and their tremendous research potential. It compiles researches conducted by experts on subsonic and supersonic wind tunnel designs, applicable for a broad range of disciplines. The book discusses various aspects of stationary and portable subsonic wind tunnel designs. It also elucidates topics related to supersonic wind tunnel and discusses a method to address fluctuating effects of fan blade rotation. This book also covers an analysis of wind tunnel applications across a multitude of engineering fields including civil, mechanical, chemical and environmental engineering.

This book is a comprehensive compilation of works of different researchers from varied parts of the world. It includes valuable experiences of the researchers with the sole objective of providing the readers (learners) with a proper knowledge of the concerned field. This book will be beneficial in evoking inspiration and enhancing the knowledge of the interested readers.

In the end, I would like to extend my heartiest thanks to the authors who worked with great determination on their chapters. I also appreciate the publisher's support in the course of the book. I would also like to deeply acknowledge my family who stood by me as a source of inspiration during the project.

Editor

Wind Tunnel Design

Design Methodology for a Quick and Low-Cost Wind Tunnel

Miguel A. González Hernández,
Ana I. Moreno López, Artur A. Jarzabek,
José M. Perales Perales, Yuliang Wu and
Sun Xiaoxiao

Additional information is available at the end of the chapter

1. Introduction

Wind tunnels are devices that enable researchers to study the flow over objects of interest, the forces acting on them and their interaction with the flow, which is nowadays playing an increasingly important role due to noise pollution. Since the very first day, wind tunnels have been used to verify aerodynamic theories and facilitate the design of aircrafts and, for a very long time, this has remained their main application. Nowadays, the aerodynamic research has expanded into other fields such as automotive industry, architecture, environment, education, etc., making low speed wind tunnel tests more important. Although the usefulness of CFD methods has improved over time, thousands of hours of wind tunnel tests (WTT) are still essential for the development of a new aircraft, wind turbine or any other design that involves complex interactions with the flow. Consequently, due to the growing interest of other branches of industry and science in low speed aerodynamics, and due to the persistent incapability of achieving accurate solutions with numerical codes, low speed wind tunnels (LSWT) are essential and irreplaceable during research and design.

A crucial characteristic of wind tunnels is the flow quality inside the test chamber and the overall performances. Three main criteria that are commonly used to define them are: maximum achievable speed, flow uniformity and turbulence level. Therefore, the design aim of a wind tunnel, in general, is to get a controlled flow in the test chamber, achieving the necessary flow performance and quality parameters.

In case of the aeronautical LSWTs, the requirements of those parameters are extremely strict, often substantially increasing the cost of facilities. But low turbulence and high uniformity in the flow are only necessary when, for example, laminar boundary layers have to be investigated. Another example of their use is aircraft engines combustion testing; this in turns requires a costly system that would purify the air in the tunnel to maintain the same air quality. Another increasingly important part of aircraft design is their noise footprint and usually the only way to test this phenomenon is in a wind tunnel.

In the automotive applications, it is obvious that the aerodynamic drag of the car is of paramount importance. Nevertheless, with the currently high level of control of this parameter and also due to imposed speed limitations, most of the efforts are directed to reduce the aerodynamic noise. The ground effect simulation is also very important, resulting in very sophisticated facilities to allow testing of both the ground effect simulation and noise production in the test section.

In architecture, due to the fact that buildings are placed on the ground and are usually of relatively low height, they are well within the atmospheric boundary layer. Therefore, the simulation of the equivalent boundary layer, in terms of average speed and turbulence level, becomes a challenging problem.

The design of the wind tunnels depends mainly on their final purpose. Apart from vertical wind tunnels and others used for specific tests (e.g. pressurised or cryogenic wind tunnels), most of the LSWTs can be categorised into two basic groups: open and closed circuit. They can be further divided into open and closed test section type.

For most applications, mainly for medium and large size wind tunnels, the typical configuration is the closed circuit and closed test chamber. Although, due to the conservation of kinetic energy of the airflow, these wind tunnels achieve the highest economic operation efficiency, they prove more difficult to design resulting from their general complexity. Hence, we will pay more attention to them in this chapter.

Apart from some early built wind tunnels for educational purposes at the UPM, since 1995 a number of LSWTs have been designed following the methodology which will be presented here. It focuses on the reduction of construction and operation costs, for a given performance and quality requirements.

The design procedure was first used for a theoretical design of a LSWT for the Spanish Consejo Superior de Deportes, which was to have a test section of $3,0 \times 2,5 \times 10,0$ m^3 with a maximum operating speed of 40 m/s. Based on this design, a 1:8 scale model was built at UPM. This scaled wind tunnel has been used for research and educational purposes.

The second time it was during the design of a LSWT for the Instituto Tecnológico y de Energías Renovables de Tenerife (ITER). That wind tunnel is in use since February 2001, operating in two configurations: medium flow quality at maximum operating speed of 57 m/s, and high flow quality at maximum operating speed of 48 m/s. For more information visit www.iter.es.

Another example of this design procedure is a LSWT for the Universidad Tecnológica de Perú, which is now routinely used for teaching purposes. This wind tunnel is now in operation for about one and a half year.

At the moment the same procedure is being utilised to design a LSWT for the Beijing Institute of Technology (BIT). This wind tunnel will be used for educational and research purposes. It will have a high quality flow, up to 50 m/s, in a test section of 1,4 x 1,0 x 2,0 m³. It will be used for typical aerodynamic tests and airfoil cascade tests (utilising the first corner of the wind tunnel circuit).

The design method to be presented in this chapter is based upon classical internal ducts design and analysis method, e.g. *Memento des pertes de charge: Coefficients de pertes de charge singulières et de pertes de charge par frottement*, I.E. Idel'cik [Eyrolles, 1986]. It also includes design assisting software such as a macro-aided Excel spreadsheet with all the complete formulation and dimensioning schemes for automatic recalculation. At the moment the best example of use of the method is the BIT-LSWT, mentioned above, as it has been defined using the latest and most reliable generation of wind tunnel design methodology.

2. Main design criteria

The general layout of the proposed wind tunnel is shown in Figure 1. The airflow circulates in the direction indicated in the test chamber (counter clockwise in the figure). Upstream of the test chamber we find the other two main components of the wind tunnel: the contraction zone and the settling chamber. The other crucial component is of course the power plant. The remainder of the components just serve the purpose of closing the circuit while minimising the pressure loss. Nevertheless, diffuser 1 and corner 1 also have an important influence on the flow quality and they are responsible for more than 50% of the total pressure loss.

The design criteria are strongly linked with the specifications and requirements and those must be in accordance with the wind tunnel applications. The building and operation costs of a wind tunnel are highly related to the specifications and these are just a consequence of the expected applications.

In the case of the so called Industrial Aerodynamics or educational applications, the require-ments related to flow quality may be relaxed, but for research and aeronautical applications the flow quality becomes very important, resulting in more expensive construction and higher operational costs.

The main specifications for a wind tunnel are the dimensions of the test section and the desired maximum operating speed. Together with this the flow quality, in terms of turbulence level and flow uniformity, must be specified in accordance with the applications. At this point it should also be defined whether all the components of the wind tunnel are going to be placed on the floor in a horizontal arrangement or in a vertical one, with only half of the circuit on the floor and the other half on top of it.

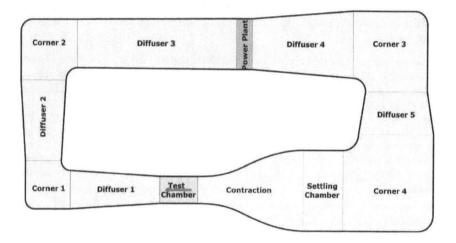

Figure 1. General layout of a closed circuit low speed wind tunnel. Figure labels indicate the part name, according to standards.

Flow quality, which is one of the main characteristics, is a result of the whole final design, and can only be verified during calibration tests. However, according to previous empirical knowledge, some rules can be followed to select adequate values of the variables that affect the associated quality parameters. The recommended values will be discussed in the sections corresponding to the Contraction, Settling Chamber, Diffusor 1 and Corner 1, which are the wind tunnel parts that have the greatest impact on the flow quality.

Once these specifications are given, it is very important to obtain on one side the overall wind tunnel dimensions to check their compatibility with the available room, and on the other side a preliminary estimation of the overall cost. The cost is mainly associated to the external shape of the wind tunnel and the power plant requirements.

For the benefit of new wind tunnel designers, a tool has been devised and implemented in an Excel spreadsheet (visit web page http://www.aero.upm.es/LSLCWT). Using this tool the designer will immediately get information about each part of the wind tunnel, the overall dimensions, the global and individual pressure loss coefficients, and the required power. This will be done according to the recommended input parameters and specification based on the intended use of the wind tunnel.

3. Wind tunnel components definition

In the following sections the design of each part will be thoroughly discussed and analysed in detail to get the best design addressing the general and particular requirements. Before dealing with each component, some general comments are given for the most important parts. In the case of the contraction zone, its design is crucial for achieving the required flow quality in the test section. In this sense, its contraction ratio, length and contour definition determine the level of uniformity in the velocity profile, as well as the necessary turbulence attenuation. It is crucial to avoid flow separation close to the walls of the contraction zone. At the stage of design, the most adequate method to verify that design meets those criteria is computational fluid dynamics (CFD).

Other important parts of the wind tunnel design worth mentioning here are the corners which incorporate turning vanes. Their aim is to reduce pressure loss and, in the case of the corner 1, possibly improve flow quality in the test section. The parameters to be considered in their design are the spacing between vanes (whether the space ought to be constant or not) and the possibility of expanding the flow (increasing the cross-section).

To complete the design process, the measurement equipment needs to be defined together with the complimentary calibration tests. Special attention needs to be devoted to the specification and selection of the balance for forces measurement, a device that is used to measure aerodynamic forces and moments on the model subjected to airflow in the test section. Since the drag force on test subjects can be very small and significant noise may be coming from the vibration of the tunnel components, such as the model stand, the true drag value may become obscured. The choice of an appropriate force balance is therefore crucial in obtaining reliable and accurate measurements.

The selection depends mainly on the nature of the tests. Wind tunnel balances can be categorized into internal and external ones. The former offers mobility since it is usually only temporarily mounted to the test section and may be used in different test sections. However, the latter has more potential in terms of data accuracy and reliability since it is tailored to a specific wind tunnel and its test section. Due to this reason, external force balances should be studied in greater depth.

3.1. Test chamber

The test chamber size must be defined according to the wind tunnel main specifications, which also include the operating speed and desired flow quality. Test chamber size and operating speed determine the maximum size of the models and the maximum achievable Reynolds number.

The cross-section shape depends on the applications. In the case of civil or industrial applications, in most of the cases, a square cross-section is recommended. In this case, the test specimens are usually bluff bodies and their equivalent frontal area should not be higher than 10% of the test chamber cross-sectional area in order to avoid the need of making non-linear blockage corrections. Accurate methods for blockage corrections are presented in Maskell (1963).

Nevertheless, a rectangular shape is also recommended for aeronautical applications. In the case of three-dimensional tests, a typical width to height ratio is 4:3; however, for two-dimensional tests a 2:5 ratio is advised in order for the boundary layer thickness in the test section to be much smaller than the model span.

Taking into account that it is sometimes necessary to place additional equipment, e.g. meas-uring instruments, supports, etc., inside the test chamber, it is convenient to maintain the operation pressure inside it equal to the local environment pressure. To fulfil this condition, it is recommended to have a small opening, approximately 1,0% of the total length of the test chamber, at the entrance of the diffuser 1.

From the point of view of the pressure loss calculation, the test chamber will be considered as a constant section duct with standard finishing surfaces. Nevertheless, in some cases, the test chamber may have slightly divergent walls, in order to compensate for the boundary layer growth. This modification may avoid the need for tail flotation correction for aircraft model tests, although it would be strictly valid only for the design Reynolds number.

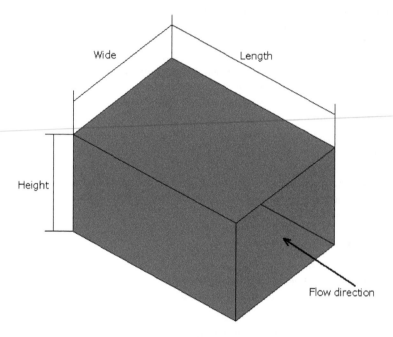

Figure 2. Layout of a constant section wind tunnel test chamber.

Figure 2 shows a design of a typical constant section test chamber. With the typical dimensions and velocities inside a wind tunnel, the flow in the test section, including the boundary layer, will be turbulent, because it is continuous along the whole wind tunnel. According to Idel´Cik (1969), the pressure loss coefficient, related to the dynamic pressure in the test section, which is considered as the reference dynamic pressure for all the calculations, is given by the expression:

$$\zeta = \lambda \cdot L \ / D_H ,$$

where L is the length of the test chamber, D_H the hydraulic diameter and λ a coefficient given by the expression:

$\lambda = 1 / (1{,}8 \cdot \log Re - 1{,}64)^2$,

where Re is the Reynolds number based on the hydraulic diameter.

3.2. Contraction

The contraction or "nozzle" is the most critical part in the design of a wind tunnel; it has the highest impact on the test chamber flow quality. Its aim is to accelerate the flow from the settling chamber to the test chamber, further reducing flow turbulence and non-uniformities in the test chamber. The flow acceleration and non-uniformity attenuations mainly depend on the so-called contraction ratio, N, between the entrance and exit section areas. Figure 3 shows a typical wind tunnel contraction.

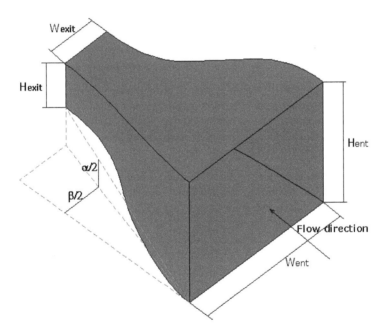

Figure 3. General layout of a three-dimensional wind tunnel contraction.

Although, due to the flow quality improvement, the contraction ratio, N, should be as large as possible, this parameter strongly influences the overall wind tunnel dimensions. Therefore, depending on the expected applications, a compromise for this parameter should be reached.

Quoting P. Bradshaw and R. Metha (1979), "The effect of a contraction on unsteady velocity variations and turbulence is more complicated: the reduction of x-component (axial) fluctuations is greater than that of transverse fluctuations. A simple analysis due to Prandtl predicts that the ratio of root-mean-square (rms) axial velocity fluctuation to mean velocity will be reduced by a factor $1/N^2$, as for mean-velocity variations, while the ratio of lateral rms fluctuations to mean velocity is reduced only by a factor of N: that is, the lateral fluctuations (in m/s, say) increase through the contraction, because of the stretching and spin-up of elementary longitudinal vortex lines. Batchelor, *The Theory of Homogeneous Turbulence*, Cambridge (1953), gives a more refined analysis, but Prandtl's results are good enough for tunnel design. The implication is that tunnel free-stream turbulence is far from isotropic. The axial-component fluctuation is easiest to measure, e.g. with a hot-wire anemometer, and is the "free-stream turbulence" value usually quoted. However, it is smaller than the others, even if it does contain a contribution from low-frequency unsteadiness of the tunnel flow as well as true turbulence."

In the case of wind tunnels for civil or industrial applications, a contractions ratio between 4,0 and 6,0 may be sufficient. With a good design of the shape, the flow turbulence and non-uniformities levels can reach the order of 2,0%, which is acceptable for many applications. Nevertheless, with one screen placed in the settling chamber those levels can be reduced up to 0,5%, which is a very reasonable value even for some aeronautical purposes.

For more demanding aeronautical, when the flow quality must be better than 0,1% in non-uniformities of the average speed and longitudinal turbulence level, and better than 0,3% in vertical and lateral turbulence level, a contraction ratio between 8,0 and 9,0 is more desirable. This ratio also allows installing 2 or 3 screens in the settling chamber to ensure the target flow quality without high pressure losses through them.

The shape of the contraction is the second characteristic to be defined. Taking into account that the contraction is rather smooth, one may think that a one-dimensional approach to the flow analysis would be adequate to determine the pressure gradient along it. Although this is right for the average values, the pressure distribution on the contraction walls has some regions with adverse pressure gradient, which may produce local boundary layer separation. When it happens, the turbulence level increases drastically, resulting in poor flow quality in the test chamber.

According to P. Bradshaw and R. Metha (1979), "The old-style contraction shape with a small radius of curvature at the wide end and a large radius at the narrow end to provide a gentle entry to the test section is not the optimum. There is a danger of boundary-layer separation at the wide end, or perturbation of the flow through the last screen. Good practice is to make the ratio of the radius of curvature to the flow width about the same at each end. However, a too large radius of curvature at the upstream end leads to slow acceleration and therefore increased rate of growth of boundary-layer thickness, so the boundary layer - if laminar as it should be in a small tunnel - may suffer from Taylor-Goertler "centrifugal" instability when the radius of curvature decreases".

According to our experience, when both of the contraction semi-angles, $\alpha/2$ and $\beta/2$ (see Figure 3), take the values in the order of 12°, the contraction has a reasonable length and a good fluid dynamic behaviour. With regard to the contour shape, following the recommendations of P. Bradshaw and R. Metha (1979), two segments of third degree polynomial curves are recommended.

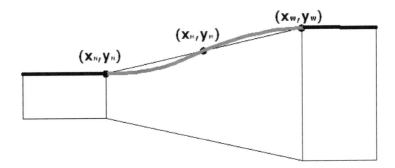

Figure 4. Fitting polynomials for contraction shape.

As indicated in Figure 4, the conditions required to define the polynomial starting at the wide end are: the coordinates (x_W, y_W), the horizontal tangential condition in that point, the point where the contour line crosses the connection strait line, usually in the 50% of such line, and the tangency with the line coming from the narrow end. For the line starting at the narrow end the initial point is (x_N, y_N), with the same horizontal tangential condition in this point, and the connection to the wide end line. Consequently, the polynomials are:

$$y = a_W + b_W \cdot x + c_W \cdot x^2 + d_W \cdot x^3,$$
$$y = a_N + b_N \cdot x + c_N \cdot x^2 + d_N \cdot x^3.$$

Imposing the condition that the connection point is in the 50%, the coordinates of that point are $[x_M, y_M] = [(x_W + x_N)/2, (y_W + y_N)/2)]$. Introducing the conditions in both polynomial equations, the two families of coefficients can be found.

According to Idel'Cik (1969), the pressure loss coefficient related to the dynamic pressure in the narrow section, is given by the expression:

$$\zeta = \left\{ \frac{\lambda}{\left[16 \cdot \sin\left(\frac{\alpha}{2}\right)\right]} \right\} \left(1 - \frac{1}{N^2}\right) + \left\{ \frac{\lambda}{\left[16 \cdot \sin\left(\frac{\beta}{2}\right)\right]} \right\} \left(1 - \frac{1}{N^2}\right),$$

where λ is defined as:

$$\lambda = 1/(1{,}8 \log Re - 1{,}64)^2.$$

The Reynolds number is based on the hydraulic diameter of the narrow section.

3.3. Settling chamber

Once the flow exits the fourth corner (see Figure 1), the uniformization process starts in the settling chamber. In the case of low-quality flow requirements, it is a simple constant section duct, which connects the exit of the corner 4 with the entrance of the contraction.

Nevertheless, when a high quality flow is required, some devices can be installed to increase the flow uniformity and to reduce the turbulence level at the entrance of the contraction (see Figure 5). The most commonly used devices are screens and honeycombs. Both devices achieve this goal by producing a relatively high total pressure loss; however, keeping in mind that the local dynamic pressure equals to $1/N^2$ of the reference dynamic pressure, such pressure loss will only be a small part of the overall one, assuming that N is large enough.

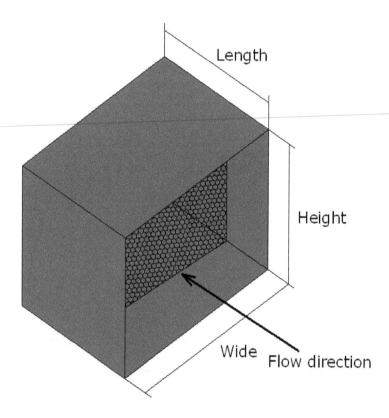

Figure 5. General layout of a settling chamber with a honeycomb layer.

Honeycomb is very efficient at reducing the lateral turbulence, as the flow pass through long and narrow pipes. Nevertheless, it introduces axial turbulence of the size equal to its diameter, which restrains the thickness of the honeycomb. The length must be at least 6 times bigger than the diameter. The pressure loss coefficient, with respect to the local dynamic pressure, is about 0,50 for a 3 mm diameter and 30 mm length honeycomb at typical settling chamber velocities and corresponding Reynolds numbers.

Although screens do not significantly influence the lateral turbulence, they are very efficient at reducing the longitudinal turbulence. In this case, the problem is that in the contraction chamber the lateral turbulence is less attenuated than the longitudinal one. As mentioned above, one screen can reduce very drastically the longitudinal turbulence level; however, using a series of 2 or 3 screens can attenuate turbulence level in two directions up to the value of 0,15%. The pressure loss coefficient, with respect to the local dynamic pressure, of an 80%-porous screen made of 0,5 mm diameter wires is about 0,40.

If a better flow quality is desired, a combination of honeycomb and screens is the most recommended solution. This configuration requires the honeycomb to be located upstream of 1 or 2 screens. In this case, the pressure loss coefficient, with respect to the local dynamic pressure, is going to be about 1,5. If the contraction ratio is 9, the impact on the total pressure loss coefficient would be about 0,02, which may represents a 10% of the total pressure loss coefficient. This implies a reduction of 5% in the maximum operating speed, for a given installed power.

The values of the pressure loss coefficients given in this section are only approximated and serve as a guideline for quick design decisions. More careful calculations are recommended for the final performance analysis following Idel'Cik's (1969) methods.

3.4. Diffusers

The main function of diffusers is to recover static pressure in order to increase the wind tunnel efficiency and, of course, to close the circuit. For that reason, and some other discussed later, it is important to maintain the flow attachment for pressure recovery efficiency. Figure 6 shows the layout of a rectangular section diffuser.

Diffuser 1 pays an important role in the test chamber flow quality. In case of flow detachment, the pressure pulsation is transmitted upstream into the test chamber, resulting in pressure and velocity non-uniformities. In addition, diffuser 1 acts as a buffer in the transmission of the pressure disturbances generated in the corner 1.

It has been proved that in order to avoid flow detachment, the maximum semi-opening angle in the diffuser has to be smaller than 3,5°. On the other hand, it is important to reduce as much as possible the dynamic pressure at the entrance of the corner 1, in order to minimise the possible pressure loss. Consequently, it is strongly recommended not to exceed the semi-opening angle limit and to design the diffuser to be as long as possible.

Diffuser 2 is a transitional duct, where the dynamic pressure is still rather large. Subsequently, the design criterion imposing a maximum value of the semi-opening angle must also be applied. The length of this diffuser cannot be chosen freely, because later it becomes restrained by the geometry of corners 3 and 4 and diffuser 5.

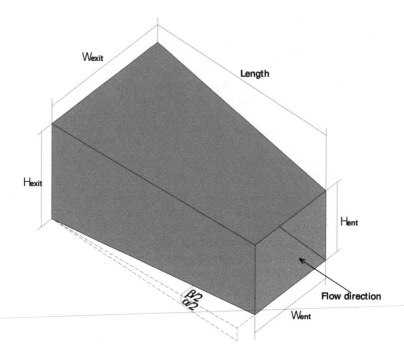

Figure 6. Rectangular section diffuser.

Diffuser 3 guides the flow to the power plant which is strongly affected by flow separation. In order to avoid it, the criterion imposing a maximum value of the semi-opening angle is maintained here as well. The cross-sectional shape may change along this diffuser because it must connect the exit of corner 2, whose shape usually resembles that of the test chamber, with the entrance of the power plant, whose shape will be discussed later.

The same can be said about diffuser 4 because pressure oscillations travel upstream and therefore may affect the power plant. Analogically to the previous case, it provides a connection between the exit of the power plant section and the corner 3, which has a cross-section shape resembling the one of the test chamber.

Diffuser 5 connects the corners 3 and 4. It is going to be very short, due to a low value of the dynamic pressure, which will allow reducing the overall wind tunnel size. This will happen mainly when the contraction ratio is high and the diffusion angle may be higher than 3,5°. It can also be used to start the adaptation between the cross-section shapes of the tests section and the power plant.

An accurate calculation of the pressure loss coefficient can be done with Idel'Cik's (1969) method. A simplified procedure, derived from the method mentioned above, is presented here to facilitate a quick estimation of such coefficient.

The pressure loss coefficient, with respect to the dynamic pressure in the narrow side of the diffuser, is given by:

$$\zeta = 4,0 \cdot \tan \alpha / 2 \sqrt{\tan \alpha/2} \cdot \left(1 - \frac{F_0}{F_1}\right)^2 + \zeta_f.$$

α being the average opening angle, F_0 the area of the narrow section, F_1 the area of the wide section and where ζ_f is defined as:

$$\zeta_f = \frac{0,02}{8 \cdot \sin \alpha / 2}\left[1 - \left(\frac{F_0}{F_1}\right)^2\right].$$

3.5. Corners

Closed circuit wind tunnels require having four corners, which are responsible for more than 50% of the total pressure loss. The most critical contribution comes from the corner 1 because it introduces about 34% of the total pressure loss. To reduce the pressure loss and to improve the flow quality at the exit, corner vanes must be added. Figure 7 shows a typical wind tunnel corner, including the geometrical parameters and the positioning of corner vanes.

The width and the height at the entrance, W_{ent} and H_{ent} respectively, are given by the previous diffuser dimensions. The height at the exit, H_{exit}, should be the same as at the entrance, but the width at the exit, W_{exit}, can be increased, giving the corner an expansion ratio, W_{exit}/W_{ent}. This parameter can have positive effects on the pressure loss coefficient of values up to approximately 1,1. However, it must be designed considering specific geometrical considerations, which will be discussed, in greater details in the general arrangement.

The corner radius is another design parameter and it is normally proportional to the width at the corner entrance. The radius will be identical for the corner vanes. Although increasing the corner radius reduces the pressure loss due to the pressure distribution on corner vanes, it increases both the losses due to friction and the overall wind tunnel dimensions. According to previous experience, it is recommended to use $0,25\ W_{ent}$ as the value of the radius for corners 1 and 2, and $0,20\ W_{ent}$ for the other two corners.

The corner vanes spacing is another important design parameter. When the number of vanes increases, the loss due to pressure decreases, but the friction increases. Equal spacing is easier to define and sufficient for all corners apart from corner 1. In this case, in order to minimise pressure loss, the spacing should be gradually increased from the inner vanes to the outer ones.

The vanes can be defined as simple curved plates, but they can also be designed as cascade airfoils, which would lead to further pressure loss reduction. In the case of low speed wind tunnels the curved plates give reasonably good results. However, corner 1 may require to further stabilise the flow and reduce the pressure loss. Flap extensions with a length equal to the vane chord, as shown in Figure 7, is a strongly recommended solution to this problem.

Other parameters, such as the arc length of the vanes or their orientation, are beyond the scope of this chapter. For more thorough approach the reader should refer to Idel′Cik (1969), Chapter 6. As mentioned above, the pressure loss reduction in the corners is very important. Therefore,

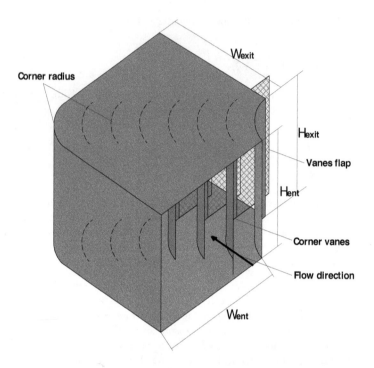

Figure 7. Scheme of a wind tunnel corner, including vanes, flaps and nomenclature.

an optimum design of these elements, at least in the case of corner 1 and 2, has a significant impact on the wind tunnel performance.

In order to allow a preliminary estimation of the pressure loss in the corners we will follow the method presented in Diagram 6.33 from Idel´Cik (1969) mentioned above. In this approach, we take an average number of vanes, $n = 1,4*S/t_1$, S being the diagonal dimension of the corner, where t_1 is the chord of the vane. The pressure loss coefficient is given by the expression:

$$\zeta = \zeta_M + 0,02 + 0,031 * \frac{r}{W_{ent}}.$$

ζ_M depends on r/W_{ent}, and its values are 0,20 and 0,17 for r/W_{ent} equal to 0,20 and 0,25, respectively. As a result, the corresponding values of ζ are 0,226 and 0,198 respectively, always with respect to the dynamic pressure at the entrance. This proves the validity of the recommendations given before with regard to the value of the curvature radius and the length of diffusor 1.

3.6. Power plant

The main aim of the power plant is to maintain the flow running inside the wind tunnel at a constant speed, compensating for all the losses and dissipation. The parameters that specify it are the pressure increment, Δp, the volumetric flow, Q, and the power, P. Once the test chamber cross-section surface, S_{TC}, and the desired operating speed, V, are fixed, and the total pressure loss coefficient, ζ, has been calculated, all those parameters can be calculated using:

$$\Delta p = \frac{1}{2}\rho \cdot V^2 \cdot \zeta$$

$$Q = V \cdot S_{TC}$$

$$P = \Delta p \cdot \frac{Q}{\eta},$$

where ρ is the operating air density and η the fan efficiency, accounting for both aerodynamic and electric motor efficiencies.

In order to reduce the cost of this part by roughly one order of magnitude, we propose to use a multi-ventilator matrix, as presented in Figure 8, instead of a more standard single ventilator power plant configuration. The arrangement of this matrix will be discussed later.

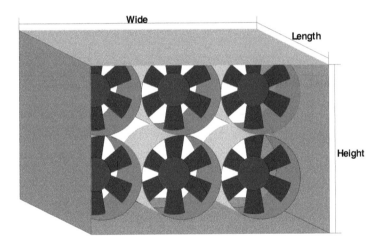

Figure 8. Layout of a multi-fan power plant.

According to our experience, for a closed circuit wind tunnel eventually including settling chamber screens or/and a honeycomb, the total pressure loss coefficient is in the range of 0,16 to 0,24. Consequently, in the case of 1,0 m² test section area and 80 m/s maximum operating speed, assuming an average value of ζ to be in the range mentioned above, and for a typical value of η equal to 0,65, the data specifying the power plant are:

Δp= 785 Pa, Q= 80 m³/s, P= 100 kW.

In this case we could use a 2,0m diameter fan specially designed for this purpose or 4 commercial fans of 1,0 m diameter, producing the same pressure increment, but with a volumetric flow of 20 m³/s each. The latter option would reduce the total cost because the fans are a standard product.

4. General design procedure

The parameters that need to be defined in order to start the overall design are:

- Test chamber dimensions: width, W_{TC}, height, H_{TC}, and length, L_{TC}. These parameters allow to compute the cross-sectional area, $S_{TC} = W_{TC} H_{TC}$, and the hydraulic diameter, $D_{TC} = 2 W_{TC} H_{TC}/(W_{TC} + H_{TC})$.

- Contraction ratio, $N \approx 5$ for low quality flow, and $N \approx 9$ for high quality flow (considering the drawbacks of choosing a higher contraction ratio, explained before).

- Maximum operating speed, V_{TC}.

According to the impact on the wind tunnel dimensions and flow quality, Table 1 shows a classification of the design variables divided into two categories: main and secondary design parameters.

Main design parameters	Secondary design parameters
Maximum operating speed, V_{TC}	Contraction semi angle, $a_c/2$
Test chamber width, W_{TC}	Settling chamber non-dimensional length, l_{SC}
Test chamber height, H_{TC}	Diffuser semi angle, $a_b/2$
Test chamber length, L_{TC}	Diffuser 1 non-dimensional length, l_{D1}
Contraction ratio, N	Corner 1 expansion ratio, e_{C1}
	Corner 1 non-dimensional radius, r_{C1}
	Corner 4 non-dimensional radius, r_{C4}
	Diffuser 5 non-dimensional length, l_{D5}
	Corner 3 non-dimensional radius, r_{C3}
	Dimension of the fan matrix, n_W, n_H
	Unitary fan diameter, D_F
	Power plant non dimensional length, l_{PP}
	Corner 2 expansion ratio, e_{C2}

Table 1. Main and secondary wind tunnel design parameters

Now, following the guidelines given above, such as the convergence angle and the contour line shape of the contraction zone, the test and contraction chamber can be fully defined. In the case when both opening angles, α and β, are the same, the contraction length, L_C, is given by the expression:

$$L_C = \frac{(\sqrt{N} - 1) \cdot W_{TC}}{2 \cdot \tan (\alpha_C / 2)}.$$

Continuing in the upstream direction, the next part to be designed is the settling chamber. The only variable to be fixed is the length, because the section is identical to the wide section of the contraction. In the case when high quality flow is required, the minimum recommended non-dimensional length based on the hydraulic diameter, l_{SC}, is 0,60. This results from the necessity to provide extra space for the honeycomb and screens. In all other cases, the non-dimensional length may be 0,50. Therefore, the length of the settling, L_{SC}, chamber is given by:

$$L_{SC} = \sqrt{N} \cdot W_{TC} \cdot l_{SC}.$$

To obtain all the data for the geometric definition of the corner 4 satisfying all the recommendations given above we only need to fix the non-dimensional radius, r_{C4}. Its length, which is the same as its width, is:

$$L_{C4} = W_{C4} = \sqrt{N} \cdot W_{TC} \cdot (1 + r_{C4}).$$

Going downstream of the test chamber, we arrive at the diffuser 1. Assuming that both semi-opening angles are 3,5°, its non-dimensional length, l_{D1}, is the only design parameter. Although it has a direct effect on the wind tunnel overall length, we must be aware that this diffuser together with corner 1 are responsible for more than 50% of the total pressure losses. According to the experience, $l_{D1} > 3$ and $l_{D1} > 4$ is recommended for low and high contraction ratio wind tunnels respectively. The length of the diffuser 1, L_{D1}, and the width in the wide end, W_{WD1}, is defined by:

$$L_{D1} = W_{TC} \cdot l_{D1}$$

$$W_{WD1} = [1 + 2 \cdot l_{D1} \cdot \tan (\alpha_{D1}/2)] \cdot W_{TC}.$$

With regard to the corner 1, once its section at the entrance is fixed (it is constrained by the exit of diffuser 1), we must define the non-dimensional radius, r_{C1}, and the expansion ratio, e_{C1}. As a result, the width at the exit, W_{EC1}, the overall length, L_{C1}, and width, W_{C1}, can be calculated using:

$$W_{EC1} = W_{WD1} \cdot e_{C1}$$

$$L_{C1} = W_{WD1} \cdot (e_{C1} + r_{C1})$$

$$W_{C1} = W_{WD1} \cdot (1 + r_{C1}).$$

Therefore, we can already formulate the overall wind tunnel length, L_{WT}, as a function of the test chamber dimensions, the contraction ratio, and other secondary design parameters:

$$L_{WT} = L_{TC} + W_{TC} \cdot \left[\frac{(\sqrt{N} - 1)}{2 \cdot \tan (\alpha_C / 2)} + \sqrt{N} \cdot l_{SC} + \sqrt{N} \cdot (1 + r_{C4}) + l_{D1} + [1 + 2 \cdot l_{D1} \cdot \tan (\alpha_{D1}/2)] \cdot (e_{C1} + r_{C1}) \right].$$

This quick calculation allows the designer to check whether the available length is sufficient to fit the wind tunnel.

Taking into account all the recommended values for the secondary design parameters, a guess value for the wind tunnel overall length, with a contraction ratio $N=9$ (high quality flow), is given by the formula:

$$L_{WT} = L_{TC} + 16 \cdot W_{TC}.$$

In the case when $N=5$ (low quality flow), the formula becomes:

$$L_{WT} = L_{TC} + 11,5 \cdot W_{TC}.$$

The designer must be aware that any modification introduced to the secondary design parameters modifies only slightly the factor that multiplies W_{TC} in the formulas above. Consequently, if the available space is insufficient, the only solution would be to modify the test chamber dimensions and/or the contraction ratio.

As we have already defined the wind tunnel length using the criterion of adequate flow quality, we can now devote our attention to designing the rest of the circuit, the so-called return circuit. The goal is not to increase its length, intending also to minimize the overall width and keeping the pressure loss as low as possible.

Keeping this in mind, the next step in the design is to make a first guess about the power plant dimensions. Following our design recommendations, a typical value for the total pressure loss coefficient of a low contraction ratio wind tunnel, excluding screens and honeycombs in the settling chamber, is 0,20, with respect to the dynamic pressure in the test chamber. This value is approximately 0,16 for a large contraction ratio wind tunnels. If screens and honeycombs were necessary, those figures could increase by about 20%.

As the power plant is placed more or less in the middle of the return duct, the area of the section will be similar to the mid-section of the contraction. Therefore, taking into account the volumetric flow, the total pressure loss, and the available fans, the decision about the type of fan and the number of them can be taken. Using this approach, the power plant would be defined, at least in the preliminary stage.

We will return now to the example we started before for the power plant section. To improve the understanding of the subject, we are going to present a case study. If the test chamber section was square and $N=5$, the mid-section of the contraction would be 1,67 x 1,67 m². This would allow us to place 4 standard fans of 0,800 m diameter each. The maximum reduction in the width size would be obtained by suppressing the diffuser 5, obtaining the wind tunnel platform shown in Figure 9. We have not defined the diffusion semi-angle in diffuser 3, but we checked afterwards that it was smaller than 3,5°. Figure 9 is just a wire scheme of the wind tunnel, made with an Excel spreadsheet, and for this reason the corners have not been rounded and are represented just as boxes.

In the case of a 4:3 ratio rectangular test chamber cross-section, the mid-section of the contraction would be 1,869x1,401 m² and for this reason we could suggest the use of 6 standard fans of 0,630 m diameter, organized in a 3x2 matrix, occupying a section of 1,890x1,260 m². Figure 10 shows the wire scheme of this new design. We can check that the diffuser 3 semi-angle is below 3,5° as well.

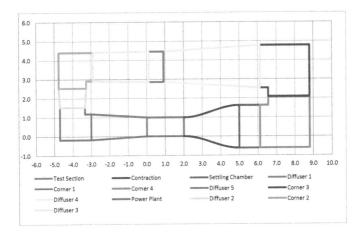

Figure 9. Non-dimensional scheme of a wind tunnel with square section test chamber and low contraction ratio, N≈5.

It is clear that the new design is slightly longer and wider, but it is because of the influence of the test chamber's width, as shown above.

Notice that in both cases corner 3 has the same shape as corner 4. Similarly, the entrance section of diffuser 4 is the same as of the power plant section, and using a diffuser semi-angle of 3,5°, this item is also well defined.

At this stage we have completely defined the wind tunnel centre line, so that we can calculate the length, L_{CL}, and width, W_{CL}, using:

$$L_{CL} = (L_{C1} - W_{EC1}/2) + L_{D1} + L_{TC} + L_C + L_{ST} + (L_{C4} - W_{ED5}/2)$$

$$W_{CL} = (W_{C4} - W_{EC4}/2) + L_{D5} + (W_{C3} - W_{ED4}/2).$$

The distance between the exit of the corner 1 and the centre of the corner 2, DC1_CC2, can be calculated through the expression (see Figure 11):

$$D_{C1\_CC2} = W_{CL} - W_{ED1}\left(r_{C1} + \frac{1}{2}\right).$$

On the other hand:

$$D_{C1\_CC2} = L_{D2} + [(W_{C2} - W_{EC2}/2)]$$

$$W_{EC2} = W_{ED2} \cdot e_{C2}$$

$$W_{C2} = W_{ED2} \cdot (r_{C2} + e_{C2})$$

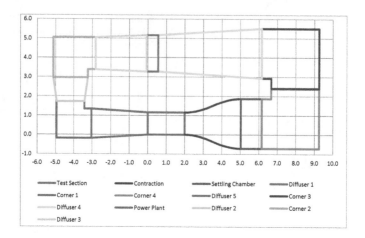

Figure 10. Non-dimensional scheme of a wind tunnel with rectangular section test chamber and low contraction ratio, $N \approx 5$.

$$W_{ED2} = W_{EC1} + 2 \cdot L_{D2} \cdot \tan\left(\alpha_{D2}/2\right).$$

Manipulating and combining those equations, we obtain:

$$L_{D2} = \frac{D_{C1\_CC2} - W_{EC1} \cdot (r_{C2} + e_{C2}/2)}{1 + 2 \cdot (r_{C2} + e_{C2}/2) \cdot \tan(\alpha_{D2}/2)}.$$

With this value, by substituting it into the previous expressions, we have all the parameters to design diffusers 2 and 3, and corner 2. Finally, it is necessary to check that the opening angles of diffuser 3 are below the limit. In case when the vertical opening angle, α, exceeds the limit, the best option is to increase the diffuser 1 length, if this is possible, because it improves flow quality and reduces pressure loss. If the wind tunnel length is in the limit, another option is to add the diffuser 5 to the original scheme. However, it will increase the overall width. When the limit of the horizontal opening angle, β, is exceeded, then the best option is to adjust the values of the expansion ratio in corners 1 and 2, because it will not change the overall dimensions.

The following case study is a wind tunnel with high contraction ratio, $N \approx 9$, and square section test chamber. In this case, the approximate area of the power plant section will be 2,000 x 2,000 m². In this case we have two compatible options to select the power plant. We can just select a matrix of 4 fans, 1,000 m diameter each. However, if the operating speed is rather high, in order to be able achieve the required pressure increment and the mass flow, we may need to use 1,250 m diameter fans. Figure 12 shows both options. Note that the overall planform is only slightly modified and the only difference is the position where the power plant is placed.

The design of the diffusers 2 and 3, and the corner 2 will be done following the same method as for the previous cases.

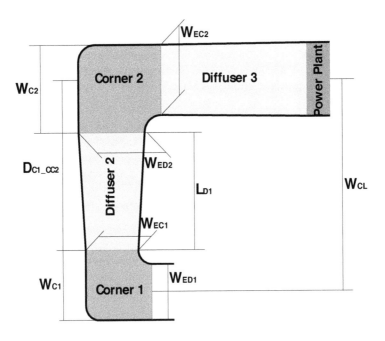

Figure 11. Scheme with the definition of the variable involving the design of diffuser 2 and 3, and corner 2.

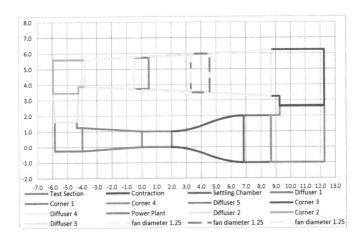

Figure 12. Non-dimensional scheme of a wind tunnel with square section test chamber and high contraction ratio, $N \approx 9$. Two different standard power plant options are presented.

5. Wind tunnel construction

One of the most important points mentioned in this chapter refers to the wind tunnel cost, intending to offer low cost design solutions. Up to now we have mentioned such modifications to the power plant, proposing a multi-fan solution instead of the traditional special purpose single fan.

The second and most important point is the wind tunnel's construction. The most common wind tunnels, including those with square or rectangular test sections, have rounded return circuits, like in the case of the NLR-LSWT. However, the return circuit of DNW wind tunnel is constructed with octagonal sections. Although the second solution is cheaper, in both cases different parts of the circuit needed to be built in factories far away from the wind tunnel location, resulting in very complicated transportation operation.

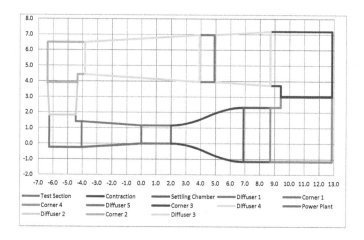

Figure 13. Non-dimensional scheme of a wind tunnel with rectangular section test chamber and large contraction ratio, N≈9.

To reduce the costs, all the walls can be constructed with flat panels, which can be made on site from wood, metal or even concrete, like in the case of ITER's wind tunnel. Figure 14 shows two wind tunnels built with wood panels and aluminium standard profile structure.

Both wind tunnels shown in Figure 14 are open circuit. The one on the left is located in the Technological Centre of the UPM in Getafe (Madrid) and its test chamber section is 1,20 x 1,00 m². Its main application is mainly research. The right one is located in the Airplane Laboratory of the Aeronautic School of the UPM. Its test chamber section is 0,80 x 1,20 m², and it is normally used for teaching purposes, although some research projects and students competitions were done there as well. Despite the fact that these tunnels are open circuit, the construction solutions can be also applied to closed circuit ones.

Figure 14. Research and education purpose wind tunnels built with wood panel and standard metallic profiles, with multi-fan power plant.

According to our experience, the manpower cost to build a wind tunnel like those defined in figures 9 to 13 could be 3 man-months for the design and 16 man-months for the construction. With these data, the cost of the complete circuit, excluding power plant, would be about 70.000,00 €. In our opinion, the cost figure is very good, considering the fact that the complete building time possibly may not exceed even 9 months.

We have more reliable data with regard to the ITER wind tunnel, built in 2000-01. The whole cost of the wind tunnel, including power plant, work shop and control room, was 150.000,00 €.

This wind tunnel was almost completely built with concrete. Figure 15 shows different stages of the construction, starting from laying the foundations to the almost final view. The small photos show the contraction, with the template used for wall finishing, and the power plant.

6. Conclusions

A method for quick design of low speed and low cost wind tunnels, either for aeronautical and/or civil applications, has been presented.

The possibility of deciding between both applications means that the method allows achieving flow quality level as good as desired.

The method also allows to the designer to get a quick and rough estimation of the overall wind tunnel size, once the main design parameters are given.

The guidelines to choose the secondary design parameters are given as well.

To address the low cost of design and construction, the use of a multi fan power plant and rectangular duct sections is proposed as well.

Figure 15. Photographic sequence of the construction of the ITER Low Speed Wind Tunnel. The top left picture shows the foundations, the top right the contraction, the bottom left the power plant and the bottom right a view from the outside almost at the end of the construction.

Nomenclature

a_i, b_i, c_i, d_i	Family of polynomial coefficients of the contraction contour shape	
CFD	Computational Fluid Dynamics	
D_{C1_CC2}	Distance between the exit of the corner 1 and the centre of the corner 2	m
D_F	Unitary fan diameter	m
D_H	Studied duct section hydraulic diameter	m
e_i	Corner 'i' expansion ratio	
F_0	Area of the diffuser's narrow section	m²
F_1	Area of the diffuser's wide section	m²
H_{ent}	Section height of the duct's entrance	m

H_{exit}	Section height of the duct´s exit	m
L	Studied duct length	m
L_i	Duct 'i' length	m
l_i	Duct 'i' non-dimensional length	
LSWT	Low Speed Wind Tunnel	
L_{WT}	Overall wind tunnel length	m
N	Contraction ratio	
n	Average number of corner vanes	
n_W, n_H	Dimensions of the fan matrix	
P	Power of the power plant	W
Q	Volumetric flow	m³/s
r	Corner radius	m
Re	Reynolds number based on the hydraulic diameter	
r_i	Corner 'i' non-dimensional radius	
S	Diagonal dimension of the corner	m
t_1	Chord of the corner vane	m
V	Operating speed at the test chamber	m/s
V_{TC}	Maximum operating speed at the test chamber	m/s
W_{CL}, L_{CL}	Wind tunnel centre line width and length	m
W_{ent}	Section width of the duct´s entrance	m
W_{exit}	Section width of the duct´s exit	m
W_{ij}, H_{ij}	Duct j width and height of the i section (wide-end, $_W$; narrow-end, $_N$; constant,)	m
WTT	Wind Tunnel Tests	
(x_N, y_N)	Narrow-end coordinates of the contraction contour shape	
(x_W, y_W)	Wide-end coordinates of the contraction contour shape	
$\alpha_i / 2$	Vertical contraction/opening semi-angle of the duct 'i'	deg
$\beta_i / 2$	Horizontal contraction/opening semi-angle of the duct 'i'	deg
Δp	Pressure increment at the power plant section	Pa
ζ	Total pressure loss coefficient	
ζ_f	Friction pressure loss coefficient	
ζ_M	Singular pressure loss coefficient of a corner	
η	Fan efficiency	
λ	Friction coefficient per non-dimensional length of the studied duct	
ρ	Operating air density	kg/m³

Acknowledgements

The authors would like to acknowledge to Instituto Tecnológico y de Energías Renovables (ITER) and to Grupo λ_3 of the UPM for their contribution.

Author details

Miguel A. González Hernández[1], Ana I. Moreno López[1], Artur A. Jarzabek[1], José M. Perales Perales[1], Yuliang Wu[2] and Sun Xiaoxiao[2]

1 Polytechnic University of Madrid, Spain

2 Beijing Institute of Technology, China

References

[1] Barlow, J. B, Rae, W. H, & Pope, A. Low-speed wind tunnel testing, John Wiley & Sons New York, (1999). rd ed.

[2] Borger, G. G. The optimization of wind tunnel contractions for the subsonic range, NASA Technical Translation / F-16899, NASA Washington, (1976).

[3] Eckert, W. T, Mort, K. W, & Jope, J. Aerodynamic design guidelines and computer program for estimation of subsonic wind tunnel performance, NASA technical note / D-8243, NASA Washington, (1976).

[4] Gorlin, S. M, & Slezinger, I. I. Wind tunnels and their instrumentation, Israel Program for Scientific Translations Jerusalem, (1966).

[5] Idel´Cik. I.E., Memento des pertes de charge: Coefficients de pertes de charge singulières et de pertes de charge par frottement, Eyrolles Editeur, Paris (1969).

[6] Maskell, E. C. A theory of the blockage effects on bluff bodies and stalled wings in a closed wind tunnel, R. & M. 3400, November, (1963).

[7] Mehta, R. D, & Bradshaw, P. Design Rules for Small Low-Speed Wind Tunnels, Aero. Journal (Royal Aeronautical Society), (1979). , 73, 443.

[8] Scheiman, J. Considerations for the installation of honeycomb and screens to reduce wind-tunnel turbulence, NASA Technical Memorandum / 81868, NASA Washington, (1981).

[9] The Royal Aeronautical Society. Wind tunnels and wind tunnel test techniques, Royal Aeronautical Society London, (1992).

Chapter 2

Design and Development of a Gas Dynamics Facility and a Supersonic Wind Tunnel

N. A. Ahmed

Additional information is available at the end of the chapter

1. Introduction

The design of a Supersonic Wind Tunnel is complex, expensive and time consuming. One of the pre-requisites of such a facility is the availability of compressed air necessary to generate the required speed.

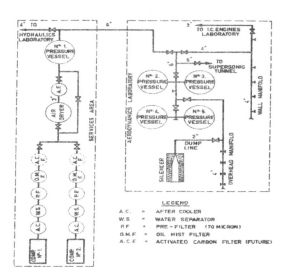

Figure 1. A Schematic of the arrangement of the Gas Dynamics Facility

In this Chapter, the design and construction of the basic gas dynamics facility (Fig. 1) is described first in Part I followed by that of a blow down type supersonic wind tunnel (Fig. 2) in Part II. The two facilities are currently in operation at the School of Mechanical and Manufacturing Engineering of the University of New South Wale.

Figure 2. A side view of the 5 ½ inch x 4 inch Supersonic Wind Tunnel

1.2. The design and construction of a Gas Dynamics facility

The Gas Dynamics facility consists of a large capacity compressed air plant that involved the installation in the Aerodynamics Laboratory of the University of New South Wales of a compressed air plant on the floor and the construction of an overhead structure of four identical 200 cubic feet capacity storage pressure vessels.

The design was initiated by setting a requirement of continuous mass flow rate of about 1 lb/sec. For continuous operation, the system pressure was set to about 100 psig. The gas dynamics facility was also required to provide air supply to a 3.5 inch x 4 inch supersonic wind tunnel, capable of operation of up to Mach 3.5.

A brief description of the gas dynamic facility is given next.

1.2.1. Compressed air plant

Compressors

From a consideration of vibration and intake resonance of the machines and also to provide significantly larger outputs per unit cost, it was decided to use rotary compressors. Two Holman RO600S screw type compressors, each rated at 600 cubic feet per minute of free air each and capable of operation separately to provide a mass flow of 0.75 lb/sec or in parallel for an output of 1.5 lb/sec were acquired. Each compressor maximum pressure ratings is 100 psig for continuous operation and 115 psig for intermittent operation such as that required for use with supersonic wind tunnels. Each unit is driven by a 150 HP 1440 RPM induction motor controlled by an auto-transformer started capable of up to 15 starts per hour. Each compressor unit was installed on 'Vulcascot' anti-vibration matting and was isolated from the discharge pipe work by means of a flexible pipe work connector. As an additional precaution, the first length of outlet pipe work to the after coolers was supported on anti-vibration matting. The result is that with both compressor operating, virtually no vibration is transmitted to the Laboratory building. A schematic of the compressed air plant is shown in Fig 1.

Control of the compressor output pressure is by either an automatic stop-start system or a constant speed uploading system operating between pre-set pressure limits. In operation, the constant speed uploading mode has been most frequently used but the original pneumatic system supplied with the compressors for this purpose proved to be unreliable. Subsequently, this was replaced with an electrical system utilising an electric control pressure gauges. This system has proved to be very satisfactory in operation and enables repeatable and readily varied settings of cut-in and cut-out pressure to be obtained with differentials as small as 2 psi.

The compressors are cooled by oil injection and lubricated by the same oil pressurised from a small pump. The cooling/lubricating oil is stored in a 40 gallon tank and cooled by an oil/water heat exchanger. The air, after compression, passes through a multi-stage reverse flow oil separator with absorbent filters so as to remove most of the oil present. Claimed oil consumption is one gallon per compressor per 400 hours of operation. Fig. 3 shows the Compressor of the gas dynamics facility.

Figure 3. The Compressor of the Gas Dynamics Facility

1.2.2. Filtration system

After compression, air from each compressor is passed through individual shell and tube after coolers, centrifugal action water separators, pre-filters and oil-mist filters. The pre-filters
consist of a centrifugal action separator combined with a 70 micron porosity, sintered
bronze filter element. The oil mist filters are a proprietary design of Daltech Engineering Incorporated, USA, and consist of a stainless wire wool pre-filter and a chemical absorption
type main element. These filter units are claimed to have a 99.4% filtration efficiency for all
oil particulates down to 0.5 microns. Both pre-filters and oil-mist filters operate with a 'colour change' principle in that, as the element becomes saturated with oil, its colour changes
from light pink to bright red. All filters, after coolers and water separators are fitted with
automatic water drains.

1.2.3. Air drier

The air drier (Fig. 4) is a proprietary designed manufactured in Australia by B.C. Johnson
Ltd. It is designed for inlet air conditions of 1.5 ib/sec at a pressure of 115 psig, a maximum
temperature of 105 ^{0} F and an ambient relative humidity at compressor inlet of 100%. The
dryer is required to provide an outlet humidity equivalent to -50 ^{0} F at atmospheric pressure
after a two hour drying cycle.

Figure 4. Air Drier System

The air dryer is of the single tower at a larger stage. Regeneration is accomplished by air pre-heated in an 80 kW electrical heater and forced through the dryer stack in a counter-flow direction by a centrifugal blower. The desiccant employed is 800 lb of activated alumina in a 2 ft diameter by 4.5 ft high bed, preceded by 80lb of buffer desiccant whose purpose is to prevent damage to the main desiccant by liquid carry over from filtration equipment. The dryer is fitted with a water coil for cooling the desiccant bed after regeneration and a felt pad and fibreglass after-filter is installed to prevent any carryover of desiccant dust into the storage vessels.

In operation, dew points of as low as -80 ^{0} F have been obtained after regeneration. With one compressor in operation, drying times of up to six hours have been achieved, although a higher final dew points than -50 ^{0} F regeneration time is about four hours for heating and four hours for cooling.

1.2.4. Air storage vessels

Four storage vessels (Fig.5), each 5 ft diameter by 11 ft 6 ins overall length and designed for a working pressure of 130 psig have been installed with a total storage capacity o f800 cubic feet as mentioned earlier were placed in an overhead structural steel support in the Aerodynamics laboratory near which a supersonic wind tunnel was built.

Figure 5. Air Storage Vessels

1.2.5. Air conditions 105 ^{0} F at atmospheric pressure

Air conditions in the storage vessels are a volume of 800 cubic feet at ambient temperature and dried to a dew point the equivalent of at least -50 ^{0} F at atmospheric pressure. For intermittent operation, maximum pressure is 125 psia. Maximum intermittent mass flow rates are in the region of 10 to 20 lb/sec to limit the system pressure loss to approximately 5 psi. Maximum continuous mass flow is 1.5 lb/sec with a pressure not exceeding 110 psia.

Because the pressure available for supersonic tunnel injection is comparatively low, care was taken in the design and piping and filtration equipment to reduce pressure losses. The overall pressure drop between compressor outlet and storage vessels has been kept between 3 and 5 psi depending upon filter condition.

1.2.6. Air distribution manifolds

The supply pipe work inter-connecting the five pressure vessels is of 6 inch inside pipe diameter pipe. There is a 6 inch branch to the supersonic tunnel. Gas dynamics rigs in the Aerodynamics Laboratory are supplied from two 4 inch pipe manifolds, one wall mounted and the other suspended from the ceiling. A four inch line, reducing to 3 inch, supply air to the Hydraulics Laboratory.

Maximum intermittent flow rates are about 10 lb/sec through the 6 inch branch supplying the supersonic tunnel and 4 lb/sec through the 4 inch manifolds. At these flow rates, the pressure drop between reservoirs and manifold outlets does not exceed about 3 psi. The hydraulics Laboratory supply system permits an intermittent flow rate of about 5 psi.

A 2 inch dump line is provided, together with a control valve and attenuating duct silencer to empty the storage vessel contents or to permit an adjustable air bled for stabilization purposes.

2. The design and construction of a blow down type supersonic wind tunnel

2.1. Design of supersonic tunnel components

Some of the details of the design of the various tunnel components are described in the following sections. The aerodynamic configuration finally selected for the tunnel is shown in Fig. 6.

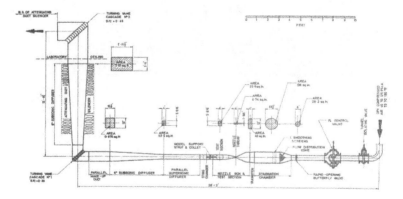

Figure 6. A Schematic of the 5 ½ inch x 4 inch Supersonic Wind Tunnel

2.1.1. Intake piping and control valves

The following requirements received attention:

• Reduction of pressure loss

• The need to supply a uniform airflow free of pulsation to the stagnation chamber

In addition to the selection of a 6 inch diameter for the tunnel intake pipe work, as mentioned in section 1.6, extra measures to reduce pressure losses and ensure flow uniformity were fitting of splitter vanes to all piping bends and tees in the final run to the tunnel. The design data of Ito (ref 29) was for this purpose.

Three valves were fitted for the flow control (Fig. 6). The first, a 6 inch gate valve, serves merely as the tunnel isolation valve and a backup shut-off valve. The second valve, downstream of the gate valve, is the stagnation pressure control valve. This is followed by the quick opening valve which is located at the inlet to the tunnel stagnation chamber.

The stagnation pressure control valve is a 6 inch double seat Fisher Governor Company valve with pneumatic cylinder actuation. The valve is of the 'Vee-pup' type which has equal percentage flow characteristics. This characteristic restricts the rate of valve spindle movement which would otherwise be necessary when the pressure drop across the valve decreases towards the end of a tunnel run. Control of the valve opening is by means of a standard 3 to 15 psi regulator located at the tunnel control panel. This regulator, acting on the positioned, supplies air at up to 100 psi to the piston of the cylinder actuator. The 6 inch valve size was selected to limit the wide open pressure drop to less than 3 psi. Preliminary design estimates indicated that the pressure drop in a 4 inch diameter control valve would have been in the region of 15 to 20 psi. The double seat valve configuration ensures reasonable symmetry in the airstream approaching the stagnation chamber and assists in reducing pressure pulsations.

The quick-opening valve is a 6 inch diameter Fisher continental rubber seat butterfly valve with pneumatic cylinder actuation and a stroking time of less than one second. It is the last component in the 6 inch line before the stagnation chamber. The position gives the most rapid possible tunnel start using standard valves. Pressure loss is about 0.1 psi. An important advantage in operation of the tunnel is gained by locating the quick-opening valve downstream of the stagnation pressure control valve as the latter can then be correctly pre-set to the required starting pressure. The valve disc position when wide open ensures flow symmetry to the stagnation chamber. The quick opening valve is actuated by a solenoid operated air valves which are, in their turn, controlled by a push button solenoid circuit on the control panel. An electrical interlock is provided so that the tunnel cannot be started until the test section access facility has been securely closed. Tunnel operation may be stopped either at the control panel or from a wandering lead and control box operated by the tunnel engineer.

Maximum Mach number in the intake pipework, excluding 'jetting' from the stagnation control valve, is of the order of 0.1. The maximum calculated pressure loss from the pressure vessels to the stagnation chamber inlet is approximately 5 psi.

The Stagnation pressure control system was deliberately chosen to be manual in orer to simplify control although a hybrid system could be incorporated at a later stage if desired. This system would consist of manual start and initial stabilisation with switch over to automatic operation once the stagnation pressure has stabilised. Some of the problems of supersonic tunnel automatic stagnation pressure control have been discussed by Pugh and Ward [1] and Conolan [2].

2.1.2. Stagnation chamber

The stagnation chamber has an inside diameter of 13.5 inches and is connected to the 6 inch inlet pipe work by a 300 conical rapid expansion. Flow stabilisation and smoothing devices consist of a conical perforated plate and flow smoothing and turbulence reduction screens (Fig. 6). A parallel settling length is provided downstream of the screens and is fitted with stagnation temperature and pressure tappings.

The mean velocity at the screens is approximately 20 ft/sec for Mach 3 operation. The chamber is also sized to permit operation down to Mach 1.5 using the same test section area when the velocity at the screens could increase to about 70 ft/sec. This remains within the range of 10 to 80 ft/sec as recommended by Pope [3].

The perforated cone has an apex angle of 900 and is manufactured from a ¼ inch plate with 3/8 inch diameter holes on 9/6 inch centres. The perforations have an open area ratio of 40%. The mean Mach number through the perforations under the worst conditions, which occur at the lowest test section Mach number, is less than 0.1. The cone is welded into the wide angle expansion. In operation, it appears to have eliminated any pressure fluctuations generated by the stagnation pressure control valve as well as assisted in 'filling' the wide angle diffuser.

The four stainless steel flow smoothing screens are of 24 mesh by 34 gauge wire and have an open area ratio of 49.5%. The screens are fixed in individual aluminium retaining ring frames. These frames are clamped together by long bolts passing through large frames attached to the rapid expansion and settling length sections of the stagnation chamber. The individual frames are spigoted together to ensure internal surface alignment and are sealed by O rings at each joint.

The parallel settling length downstream of the screen is 1500 screen wire diameters long, or a length of approximately 18 inches.

A two dimensional contraction and section change transition region is provided at the downstream end of the settling chamber. This region has a rectangular outlet area of 12 inch x 4 inch and a circular 13.25 inch diameter inlet section. The area contraction ratio is 2.9:1. A further two-dimensional contraction, of ratio 10:1, is built into the nozzle blocks to contract the airstream to a sonic throat 1.185 inch high by 4 inches wide. The method of contraction design presented by Gibbings [4] is recommended. This method is also applicable to contractions in which there is an appreciable axial trigger between the plan and elevation profiles.

In operation, the stagnation chamber provides a steady pressure with accuracy of control of 1% or better. Estimated pressure loss for the stagnation chamber flow smoothing devices was 2 psi approximately, making a total estimated loss between pressure vessels and stagnation pressure measurement station of nearly 7 psi. In operation, this pressure loss varies between 7 -10 psi.

2.1.3. Nozzle box and test section

The nozzle box is of conventional construction and is manufactured from steel plate with internal surfaces ground after welding and stress relieving. Heavy stiffening ribs prevent deformation under pressure forces, particularly in the throat region. Dowels are fitted to ensure accurate and repeatable alignment of adjacent parts. Circumferential 'O' ring seals are provided at each end of the nozzle box. One side wall opens downwards on hinges to facilitate nozzle block changes. The complete assembly of nozzle box and settling length section of the stagnation chamber can be moved on rollers to permit easy screen removal. The rollers are brought into operation by four jack screws. An axial movement of 3 inches in the downstream direction is possible.

There are circular Schlieren window positions in the nozzle box walls, one pair at the throat and one pair at the test section. The windows have a clear diameter of 7.5 inches and thickness of 1 inch and are held in a sub-frame which is fixed to the tunnel by a clamping ring. This arrangement permits easy removal and rotation of each window. Rotation of the window assembly permits selection of the optimum orientation for optical characteristics of the glass fitted. This allows the use of cheap and comparatively low grade plate glass. The glass is sealed to the sub-frame with Dow Corning Silastic 732 RTV silicon rubber compound. A special jig has been developed for window assembly which ensures that the glass is completely 'floating' in Silastic and is also flush to within 0.001 inches with the frame edges. A set of high quality glass windows obtained from Optical Works, UK are also available for specially sensitive Schlieren applications.

The supersonic nozzle blocks are manufactured from extruded AA28S aluminium alloy. The contours were generated by a programmed 'Hydroptic jig borer and were finally hand finished to remove machining marks. Each block is fitted with a continuous, circular cross section rubber seal which runs as close to the contoured surface as is possible using straight line approximations. The nozzle blocks are held in the nozzle box with the bolts passing through the top and bottom walls of the box into barrel nuts inside each block. Location is by integral machined pads on the basis of each block. It is now realised that the provision of separate ground pads on the base of each nozzle block configuration would have permitted the fitting of permanent shims clamped between pad and block, thereby simplifying the fitting and accurate setting up of each nozzle block configuration within the nozzle box.

The nozzle block co-ordinates are those derived by McCabe [5] for operation at Mach 3 in a nominal 5.5 inch x 5.5 inch test section. The design method used by McCabe divides the nozzle inviscid core flow into five main regions, as follows:

• A subsonic contraction

- The high subsonic, low supersonic throat region
- An initial expansion region wherein the contour slope increases to its maximum value
- The straightening or 'Busemann Region'
- The parallel flow or test section region

McCabe also divides each of the 3rd and 4th regions into further regions so as to take advantage of more precise computational methods and eliminate discontinuities in nozzle curvature. The nozzle contour boundary layer corrections are based on the data of Sibulkin [6] for the throat region, with corrections on the contoured and flat side walls obtained from the data of Rogers and Davis [6]. Because the nozzle contours so derived did not have zero slope at the test section location, a cubic curve was fitted to permit smooth transition to the parallel block region downstream of the test rhombus. The boundary conditions were continuity of ordinate, slope and second derivative at the upstream and zero slope at the downstream end. Before feeding to the jig borer, the complete set of nozzle block co-ordinates was smoothed by computer using a 6th order polynomial. Data on nozzle profile design may be obtained from Refs 8-15.

2.1.4. Diffusers and model support system

Downstream of the nozzle box are a string chamber, supersonic diffuser, first stage subsonic diffuser, parallel make up duct and a set of cascade turning vanes. The complete assembly downstream of the nozzle box to the corner cascade is mounted on a flanged wheel and rail system and may be moved 18 inches in the downstream direction away from the test section. This allows easy access for model changes. The face of the sting chamber is aligned with, and closed against, the nozzle box pressure seal by four dowel pins and tapered locking wedges. Four cam clamps are used to close the pressure seal between the cascade corner and the vertical second stage subsonic diffuser.

The sting chamber can be removed entirely, if required. It is fitted with a pair of circular openings of the same diameter as those in the nozzle box side walls and, can, therefore, accept the interchangeable Schlieren windows or metal window blanks. A side mounted sting may be fitted to either of the side window blanks. A vertical, full span, wedge nose strut is also in use as a model sting support.

The optimum design of a supersonic diffuser for a small wind tunnel presents a difficult problem as it must operate over a wide range of inlet flow conditions and Reynolds numbers. Moreover, at Mach numbers in excess of 2, optimum starting and running diffuser geometries diverge significantly. It is assumed in the following discussion that the supersonic diffuser is followed by a subsonic diffuser of small divergence angle.

In general, the airstream entering the diffuser is highly turbulent because it contains the wake flow from the test section model and its support system. Moreover, diffuser performance is influenced by boundary layer effects which are a function of tunnel Reynolds number.

Supersonic diffusers for small wind tunnels may be one, or a combination of the following types:

- Fixed contraction with constant area second throat

- Variable contraction with variable area second throat

- Constant area duct

- Oblique shock diffuser with centre body

The diffuser throat must be sized must be to 'swallow' a normal shock when starting at the design Mach number in a second throat type of diffuser. Faro [11] obtained the theoretical minimum starting contraction ration required, together with experimental points from several sources. For Mach numbers in excess of 2, the experimental values are lower than the theoretical values. It is suspected that this is due the starting shock passing through the second throat before the design Mach number has been reached. The experimental data of Lukasiewicz [16] indicates that at the area ratio of $A_2/A_1 = 0.7$, which is required for stating at Mach 3, the starting pressure ratio requirement of a fixed geometry second throat diffuser, might be 0 to 30% less than that required by an optimum length constant area diffuser. If a variable geometry second throat diffuser is used, significant reductions in optimum running contraction ratio may be obtained particularly at Mach numbers in excess of 2.

Variable geometry diffusers, set at optimum running contraction ratios after starting, enable reductions in running pressure ratio of up to about 45% when compared with optimum length constant area ducts [17]. This reduction is achieved, however, at a considerable increase in complexity of construction and operation when compared with the simple constant area diffuser.

The advantages of the fixed and variable area second throat diffusers over the constant area diffuser are less certain when allowance is made for the presence of the model and its support system within the test section. Faro [11] states that in nearly all cases the presence of a model reduces supersonic diffuser efficiency. This is confirmed by De Leo and Huerta [17] who found that a 5% blockage model increased starting and running pressure ratios by approximately 7 and 13 % respectively for optimum empty tunnel diffuser geometry and Mach numbers of 2.5 and 3.4. However, experiments have also shown that the presence of a model tends to reduce tunnel instability through stabilisation of the diffuser shock system. This appears to be due to interaction between the wall reflections of the model bow wave and the diffuser shock system. In summary, therefore, it seems that the second throat diffuser types do not offer significant advantages over the constant area duct for small supersonic tunnels. This conclusion does not apply to very large intermittent type tunnels, however, as then a reduction in running pressure ratio could lead to significant increases in tunnel run time for a given pressure storage capacity or reduction in capital cost of the storage system for a given run time.

The supersonic diffuser of the University of New South Wales is a constant area duct system although the sting strut, when used probably acts as a centre body with oblique shock diffusion. Although such a constant diffuser is a dissipative system, it does have the considerable advantage of being stable over a wide range of inlet flow conditions and of being simple to construct and maintain.

The only design decision required for a constant area supersonic diffuser is the optimum length of duct. In such a diffuser the normal shock appears as a shock system which is strongly affected by the state of the boundary layer. Experimental dat quoted by Lukasiewcz [16] confirms that for best efficiency the shock compression process should be completed in the constant area duct and not in the divergent subsonic diffuser downstream of the supersonic diffuser. Faro [11] illustrates the gain in isentropic efficiency with increasing length to height ratio for a constant area duct at Mach 2. The significant reduction in operating pressure ratio with increasing length of parallel duct may be seen in curves plotted for length to height ratios of 0, 2 and 7. Further design data for constant area diffusers is demonstrated by Faro [11]. Two points are noted in connection with his work:

- The Mach number M_{av} is the average Mach number at the supersonic diffuser inlet and would be less than the test section Mach number because of the presence of the model and its support system and boundary layer growth between test section and diffuser.

- The effect of the free stream Reynolds number is not accounted for. Some indication of the reduction in the length of the shock compression system at high Reynolds numbers may be obtained from [16]. This data is applicable to a Mach number of approximately 2.

Faro [26] indicates that a single wedge such as the leading edge of a sting support strut may be used to provide an oblique shock system which will improve diffuser efficiency over the simple normal shock case. The benefits for this type of device are limited, however, is that high efficiency can only be obtained with a large number of oblique shocks which in turn implies design for a specific Mach number and thus a narrow range of off-design conditions. The simple strut type oblique shock generator gives moderate efficiency gains over a wider range of Mach numbers.

To summarise, little data for the design of constant area supersonic diffusers or for the effect of a model and strut system on diffuser efficiency can be found. The available information suggests that:

- The shock system compression process should be completed within the parallel diffuser duct for best efficiency

- The optimum length of parallel duct required to complete the compression process is a strong function of Mach number and Reynolds number. This length is probably within the 5 to 12 diffuser heights of Mach numbers of 1.5 to 3.5 and Reynolds numbers of 2×10^5 to 6×10^6. Design data for supersonic diffusers may be obtained from Refs 16-22].

The supersonic diffuser of the University of New South Wales tunnel is a parallel wall rectangular duct fabricated from 4 inch x 1 inch extruded aluminium bar top and bottom walls and 0.5 inch aluminium plate side walls. The top and bottom walls may be easily replaced with a set of contoured blocks so as to provide a fixed area second throat, if so desired. The parallel diffuser length is 8.4 diffuser heights from the rear of the model support strut and 11.4 heights from the end of the supersonic nozzle with model support system removed. A removable parallel subsonic make-up duct permits the fitting of an additional 4.3 diffuser heights. The window blanks alongside the model support strut also allow for experiments involving 'de-blocking' of the area around the strut.

The design of the subsonic diffuser was straightforward. The data of Lukasiewicz [16] indicates that, in the Mach number range 0.4 to 0.9, total pressure recovery is virtually constant at about 0.88 for open ducts without models and that the diffuser divergence angle should be less than 7^0. Data on subsonic diffuser design is available from Refs 14 and 16.

There are two stages of subsonic diffusion separated by the corner cascade (Fig. 6). The first stage of subsonic diffusion is separated by the corner cascade (Fig. 6). The first stage has an area ratio of 5.6 and divergence angle of 60. Maximum Mach number at the subsonic diffuser exit is approximately 0.13. The diffuser is constructed from 3/16 inch steel plate reinforced at 6 inch x 3 inch centres with 1 x 0.25 inch flat bars on edge.

The second stage subsonic diffuser has a 60 divergence angle and 3:1 area ratio. Maximum Mach number at exist to this diffuser is approximately 0.04. The diffuser is manufactured from plywood and incorporates part of the tunnel silencing system. The corner cascade utilises sheet metal circular arc turning vanes.

2.1.5. Silencer

Preliminary investigations on an existing M3.5, 4 inch diameter conical nozzle indicated that the noise level for an unsilenced tunnel would be unacceptably high at about 120 db in the frequency band of 100 to 2000 Hz. Accordingly, a silencer was designed for the supersonic tunnel to the following requirements:

- Noise reduction to about 80 dB in the 100 to 2000 Hz band

- Low pressure loss

- Ease of construction and low cost

After investigation, an attenuating duct design was chosen as best fulfilling these requirements. This type of silencer requires an absorbent material as dense as possible with a thickness of 2 inch to 12 inches to absorb the low frequency noise below 500 Hz. Attenuation at the lower frequencies is increased considerably by the use of a perforated duct facing material having about 3 to 10% open area perforations. Low frequency attenuation is further assisted by providing airspace behind the absorbent material and increasing the amount of absorbent around the duct periphery. When compared with splitter type duct attenuators, low frequency attenuation can be improved by arranging a given amount of attenuation material such that it forms thick layers. This latter arrangement gives a lower peak but better average attenuation over the 100 to 1000 Hz frequency band. Design information can be found in the literature [23-29].

The silencer for the University of New South Wales tunnel is constructed in two sections: the first of which is built around the second stage subsonic diffuser. The first section comprises 6 to 12 inch thickness of polythene wrapped rockwool batts and loose rockwool fill around around all four sides of the diffuser. The rockwool density varies from 4 to 6 lb/cubic feet for the loose fill. The duct interline is surfaced with 3/16 inch thick perforated plywood and the outside of the silencer is sealed with 1 inch thick, exterior quality waterproof plywood. Both internal and external surfacing materials are heavily glued, screwed and nailed to substantial connecting framing. The second section of the silencer, which is 16 ft long is a rectangular duct lined on two sides with 6 inch thickness of rockwool batts backed by a 3 inch airspace. The remaining two sides of this duct are 1 inch thick exterior plywood. Other constructional details are similar to those of the first section silencer. The second diffuser section is run in the laboratory ceiling space and is supported from the roof structure on 'Silentbloc' vibration isolators.

Initial tests on completion of the tunnel indicated a large direct sound transmission through the walls of the first stage subsonic diffuser. This was found to be caused by high frequency resonance of the 3/16 inch thick flat steel plate walls. The vibration was almost completely eliminated and the noise level reduced by decreasing the spacing of the existing 1 inch x 0.25 inch stiffening bars from approximately 12 inch x 6 inch to 6 inch x 3 inch centres as described in section 4.4.

In the final form, the silencer has reduced the noise level in the vicinity of the tunnel to about 75 to 90 dB, for the 100 to 2000 Hz band, depending to some extent upon the operating stagnation pressure. It is estimated that the duct silencer provides an attenuation of about 2 to 3 dB per foot of length in the frequency range 100 to 1000 Hz.

2.1.6. Instrumentation

The tunnel stop-start system has been briefly described in section 2.1.4.

Tunnel stagnation pressure is read on 0.15% accuracy, temperature compensated, absolute pressure 'Heise' test gauge and recorded by a pressure transducer having 0.1% combined non-linearity and hysteresis. The transducer output can be displayed directly in psia on an 11 inch 'Honewell' strip chart recorder. The control panel is provided with an electrically actuated pneumatic calibration circuit which connects the stagnation pressure transducer and test gauge in a closed system. This circuit has an electrical override if the tunnel is started.

Stagnation temperature instrumentation consists of an exposed-junction 'BLH' micro-miniature thermocouple connected to an 11 inch strip chart recorder and reading directly in 0F. Bothe stagnation temperature and pressure recorders contain electrically operated chart speed-up facilities which automatically increase the chart speed by a factor of 60:1 when the tunnel run is started. A typical speed change is from 10 inches per hour to 10 inches per minute. Both chart recorders are provided with event markers which are connected into the tunnel timing circuit. The circuit operates an electrically actuated second timer which is controlled from a timer switch in the remote control box on a wandering lead. The box also contains the tunnel stop switch and a pressure clamp switch. The event markers are automatically actuated at the start and stop of a timing run. The wandering lead control box enables one man to control the run and monitor Schlieren and instrument read out.

Data acquisition is by conventional multi-manometers, pressure gauges or a range of flush diaphragm transducers. For calibration, the data transducers can be connected into a closed system with a 'heise' stagnation pressure gauge. This only requires operation of a control panel push-button. Read out equipment for the transducers are EAI and Solartron multi-channel data loggers and a 6 channel pen recorder.

Tunnel air flow calibration equipment has been designed and built in accordance with the data published by Anderson [16]. Air humidity is measured with a Casella-Alnor dew point meter which can measure dew points to -80⁰F with an accuracy of 3⁰F.

A two-mirror, parallel path, 96 inch focal length, 7.5 inch aperture Schlieren system is currently in use with the tunnel and associated gas dynamics rigs. A second 60 inch focal length system is being assembled. Both systems are portable, freely adjustable and provided with heavy bases. Photographic facilities include a 5 inch x 4 inch plate camera specially adapted for daylight use with the Schlieren system along with a high speed drum camera and a cine camera oprating at framing speeds up to 20,000 frames per second.

2.2. Operational problems

The gas dynamics and supersonic tunnel facilities have proved to be simple to operate, reliable and comparatively trouble-free. However, there have been two operational problems which may of interest, and they are described below.

- Oil mist carry over: A slight oil deposition occurs on the tunnel test section windows after several runs indicating that the oil filtration is not completely effective. It is likely that this results from the fact that the oil mist filters have their highest removal efficiency for aerosol having particulate sizes down to 0.5 micron. It is possible that a very fine oil aerosol is passing through the filters with particulates in the size range 0.1 to 0.001 micron. A solution to the problem would be the use of a hydrocarbon selective absorbent which has a high removal efficiency in the size range where Brownonian motion predominates. For this reason, a set of activated carbon filters was manufactured and installed.

- Water Condensation in Pipelines: Operation of the gas dynamics facility without the air dryer is currently necessary for certain types of tests. As a result, liquid water condenses in the pipelines and pressure vessels and ultimately results in small rust flakes periodically discharging into a stagnation chamber. The solution to this problem appears to be use only dry air for all operations. The existing single dryer tower does not have sufficient drying capacity for some rig work where run durations of up to eight hours are required. Consequently, installation of the second dryer tower appears to be necessary, although much of this type of running can be done at higher outlet dew points than the supersonic tunnel operation, thus permitting longer run times before regeneration. It also appears desirable to introduce some form of direct reading humidity meter which could also be used to operate a warning system. As an interim measure to combat water collecting in the pipelines and fittings, all low points of the system have been provided with water drain valves.

3. Conclusion

A compressed air plant, providing 1.5 lb/sec of dry air at 100 to 115 psig and having a storage capacity of 1000 cubic feet, has been engineered and built. The system has been operating satisfactorily apart from an oil mist problem for which corrective measures are being investigated.

A supersonic blowdown wind tunnel, using air from the compressed air plant and exhausting to atmosphere, has also been built to simple, conventional design principles. Nozzle blocks for Mach 3 and parallel duct supersonic diffuser has been installed. Although stagnation pressure control is manual, the tunnel is designed for operation by only one man. Run time varies from 20 to 60 seconds and test section Reynolds numbers of about 10^6 per inch may be obtained.

These facilities along with subsonic wind tunnel facilities form the basis of aerodynamic, research and development works at the University of New South Wales [31-89]

Acknowledgements

The Author wishes to gratefully acknowledge the hard works and dedication of Barry Motson and the late Associate Professor Archer in the Design of the Gas Dynamics facility and the Supersonic Wind Tunnel.

Author details

N. A. Ahmed*

Aerospace Engineering, School of Mechanical and Manufacturing Engineering, University of New South Wales, Sydney, NSW, Australia

References

[1] Pugh, P. G., & Ward, L. C. (1966). Notes on the Automatic Control of a Blowdown Wind Tunnel'. *NPL Aero Report*, 1215.

[2] Conolan, K. G. (1964). Control equipment for the ARL Hypersonic Wind Tunnel',. *ARL TM*, 196.

[3] Pope, A., & Goin, K. L. (1984). High speed wind tunnel testing',. Wiley, NY.

[4] Gibbings, J. C. (1965). A Note on the Combination of a Contraction with a Supersonic Nozzle for Wind Tunnel',. *RAE TR*, 65066.

[5] Mc Cabe, A. (1964). Design of Supersonic Nozzle',. ARC Fluid Motion Sub-Committee, ARC 25716, FM 3433, March.

[6] Sibulkin, M. (1956). Heat Transfer to an Incompressible Turbulent Boundary Layer and Estimation of Heat Transfer Coeffcients at Supersonic Nozzle Throats',. *JAS,,* 23(2), 162.

[7] Rogers, E. W. E., & Davis, B. M. (1956). A note on Turbulent Boundary Layer Allowances in Supersonic Nozzle Design. *ARC CP*, 333.

[8] Dodd, K. N., et al. (1964). Calculations for the Design of Nozzles'. *RAE TR*, 64021.

[9] Tucker, M. (1951). Approximate Calculation of Turbulent Boundary Layer Development in Compressible Flow. *NACA TN*, 2337.

[10] Beckwith, J. E., & Moore, J. A. (1955). An Accurate and Rapid Method for the Design of Supersonic Nozzles. *NACA TN*, 3322.

[11] Faro, D. V., & (ed, . (1964). Handbook of Supersonic Aerodynamics', NAVWEPS Rep 1488, 6 sec 17, Ducts,. *Nozzles and Diffusers',*.

[12] Pierce, D. (1965). A Simple Flexible Supersonic Wind Tunnel Nozzle for the Rapid and Accurate Variation of Flow Mach Number. *RAE TR*, 65280.

[13] Liepman, H. P. (1955). Analytic Method for the Design of 2D Asymmeric Nozzles',. *J of Aero Sc.,,* 701-709.

[14] Culley, M. (1966). The ARL Supersonic Propulsion Wind Tunnel: Redesign of the Nozzle Profile and preliminary Calibration',. *ARL ME* [282].

[15] Daniels, W. (1956). Design and development of the North American Aviation Transonic Wind Tunnel',. *AGARD* [67].

[16] Lukasiewicz, J. (1953). Diffusers for Supersonic Wind Tunnels',. *J of Aero Sci,,* 20(9), 617-626.

[17] De Leo, R., & Huerta, J. (1956). The Influence of Adjustable Diffuser Configurations on Minimum Required Starting and Operating Pressure Ratios for a Supersonic Wind Tunnel',. *Proc of 5th Biennial Tech Conf. Uni of Minnesota, Research Report* [137].

[18] Tucker, N. B. (1956). Data on Temperature Stabilisation and Diffuser Performance of the 5 inch x 5 inch Pilot Blowdown Wind Tunnel,. *AGARD* [92].

[19] Lucasiewicz, J. (1964). Pressure Measurement in Wind Tunnel S3,. *WRE TN HSA*, 132.

[20] Ferri, , & Bogdonoff, S. F. (1956). Design and Optimisation of Intermittent Supersonic Boundary layer and Estimation of Heat Transfer Coefficients at Supersonic Nozzle Throats',. *JAS*, 23(23), 62.

[21] Leavy, L. E. (1956). A supersonic Wind Tunnel for Mach Numbers up to 3. *AGARD* [70].

[22] Herman, R. (1956). A Basic Theorem Concerning Pressure Recovery of Symmetric Supersonic Diffusers',. *Proc of 5th Biennial Tech Conf. Uni of Minnesota, Research Report* [137].

[23] Beranek, L. L. (1954). Acoustics',. McGraw-Hill, NY.

[24] Ahmed, N. A. Design Features of the Low Speed 4 ft x 3 ft Return Circuit Wind Tunnel'.

[25] Callaway, D., & Ramer, L. G. (1952). The use of Perforated Facings in Designing Low-Frequency Resonant Absorbers. *J of Acoustical Soc of America,,* 24(3), 309-312.

[26] Beranek, L. L., Labate, S., & Ingrad, U. (1955). Noise Control for NACA Supersonic Wind Tunnel. *J of Acoustical Soc of America,,* 27(1), 85-98.

[27] Myer, E., Miechel, F., & Kurtze, G. (1958). Experiments on the Influence of Flow on Sound Attenuation in Absorbing Ducts',. *J of Acoustical Soc of America,,* 30(3), 165-174.

[28] Ingard, U., & Bolt, R. H. (1951). Absorption Characteristics of Acoustical Material with Perforated Facing. *J of Acoustical Soc of America,,* 21(2), 533.

[29] Beranek, L. L. (1960). Noise Reduction'. McGraw Hill, NY,.

[30] Anderson, A. (1963). Flow Characteristics of a 12 Inch Intermittent Supersonic Tunnel',. *AEDC Tech Doc Report,* AEDC-TDR-, 63-203.

[31] Ahmed, N. A. (2010). Wind driven Natural-Solar/Electric Hybrid Ventilators',. in ', *Wind Power,* ', Section D: The Environmental Issues",, Chapter 21, edited by S. M. Muyeen Kitami, published by In-Tech Organization, Austria,, 978-9-53761-981-7.

[32] Lienand, J., & Ahmed, N. A. (2011). Wind driven ventilation for enhanced indoor air quality',. invited Chapter, in 'Chemistry, Emission, Control, Radiaoactive Pollution and Indoor Air Quality', edited by Nicholas A Mazzeo, published by In-Tech Organization, Austria, 978-9-53307-570-9

[33] Findanis, N., & Ahmed, N. A. (2011). Wind tunnel 'concept of proof' investigations in the development of novel fluid mechanical methodologies and devices',. invited Chapter, in ', *Wind Tunnels and Experimental Fluid Dynamics Research',,* edited by J. C. Lerner and U. Boldes, published by In-Tech Organization, Austria,, 978-9-53307-623-2.

[34] Ahmed, N. A., Elder, R. L., Foster, C. P., & Jones, J. D. C. (1987). A Novel 3D Laser Anemometer for Boundary Layer Studies',. ASME Conf, Boston, USA, 15th December., Also in the, *3rd International Symposium on Laser Anemometry,* edited by A Dybs et al, ASME, The Fluids Engineering Division,, 55, 175-117.

[35] Ahmed, N. A., Elder, R. L., Foster, C. P., & Jones, J. D. C. (1990). Miniature Laser Anemometer for 3D Measurements. *J of Measurement Sc Technol,,* 1, 272-276.

[36] Ahmed, N. A., Elder, R. L., Foster, C. P., & Jones, J. D. C. (1990). Miniature Laser Anemometer for 3D Measurements. *Engineering Optics,,* 3(2), 191-196.

[37] Ahmed, N. A., Elder, R. L., Foster, C. P., & Jones, J. D. C. (1991). Laser Anemometry in Turbomachines. *IMechE Proc, Part G, J of Aerospace Engineering,*, 205, 1-12.

[38] Ahmed, N. A., Hamid, S., Elder, R. L., Foster, C. P., Jones, J. D. C., & Tatum, R. (1992). Fibre Optic Laser Anemometry for Turbo machinery Applications', *Optics and Lasers in Engineering,*, 15(2-3), 193-205.

[39] Ahmed, N. A., & Elder, R. L. (2000). Flow Behaviour in a High Speed Centrifugal Impeller Passage under Design and Off-design Operating Conditions', *Fluids and Thermal Engineering, JSME International*, 43(1), 22-28.

[40] Simpson, R. G., Ahmed, N. A., & Archer, R. D. (2000). Improvement of a Wing Performance using Coanda Tip Jets', *AIAA Journal of Aircraft,*, 37(1), 183-184.

[41] Gatto, A., Ahmed, N. A., & Archer, R. D. (2000). Investigation of the Upstream End Effect of the Flow Characteristics of a Yawed Circular Cylinder', *The RAeS Aeronautical Journal,*, 104(1033), 253-256, pp125-128.

[42] Gatto, A., Ahmed, N. A., & Archer, R. D. (2000). Surface Roughness and Free stream Turbulence Effects on the Surface Pressure over a Yawed Circular Cylinder', *AIAA Journal of Aircraft,*, 38(9), 1765-1767.

[43] Ahmed, N. A., & Archer, R. D. (2001). Performance Improvement of a Bi-plane with Endplates', *AIAA Journal of Aircraft,*, 38(2), 398-400.

[44] Gatto, A., Byrne, K. P., Ahmed, N. A., & Archer, R. D. (2001). Pressure Measurements over a Cylinder in Crossflow using Plastic Tubing', *Experiments in Fluids,*, 30(1), 43-46.

[45] Ahmed, N. A., & Archer, R. D. (2001). Post-Stall Behaviour of A Wing under Externally Imposed Sound', *AIAA Journal of Aircraft,*, 38(5), 961-963.

[46] Ahmed, N. A., & Archer, R. D. (2002). Testing of a Highly Loaded Horizonatal Axis Wind Turbines designed for Optimum Performance', *International Journal of Renewable Energy,*, 25(4), 613-618.

[47] Simpson, R. G., Ahmed, N. A., & Archer, R. D. (2002). Near Field Study of Vortex Attenuation using Wing Tip Blowing', *The Aeronautical Journal*, 102.

[48] Ahmed, N. A., & Goonaratne, J. (2002). Lift augmentation of a low aspect ratio thick wing at a very low angle of incidence operating in ground effect", *AIAA Journal of Aircraft,*, 39(2).

[49] Ahmed, N. A. (2002). Implementation of a momentum integral technique for total drag measurement', *International Journal of Mechanical Engineering and Education,*, 30(4).

[50] Pissasale, A., & Ahmed, N. A. (2002). Theoretical calibration of a five hole probe for highly three dimensional flow', *International Journal of Measurement Science and Technology,*, 13, 1100-1107.

[51] Pissasale, A., & Ahmed, N. A. (2002). A novel method of extending the calibration range of five hole probe for highly three dimensional flows",. *Journal of Flow Measurement and Instrumentation,,* 13(1-2), 23-30.

[52] Ahmed, N. A., & Wagner, D. J. (2003). Vortex shedding and transition frequencies associated with flow around a circular cylinder",. *AIAA Journal,* 41(3), 542-544.

[53] Rashid, D. H., Ahmed, N. A., & Archer, R. D. (2003). Study of aerodynamic forces on rotating wind driven ventilator',. *Wind Engineering,* 27(1), 63-72.

[54] Ahmed, N. A. (2003). An acoustic energy concept for the design of a flow meter',. *International Journal of Vibration and Acoustics,,* 8(1), 52-58.

[55] Pissasale, A., & Ahmed, N. A. (2003). Examining the effect of flow reversal on seven-hole probe measurements',. *AIAA Journal,,* 41(12), 2460-2467.

[56] Pissasale, A., & Ahmed, N. A. (2004). Development of a functional relationship between port pressures and flow properties for the calibration and application of multi-hole probes to highly three-dimensional flows',. *Experiments in Fluids,,* 36(3), 422-436.

[57] Ahmed, N. A. (2004). Demonstration of the significance and limitations of thin airfoil theory in the study of aerodynamic characteristics of an airfoil",. *IJMEE,,* 34(2), 271-282.

[58] Ahmed, N. A. (2006). Investigation of dominant frequencies in the transition Reynolds number range of flow around a circular cylinder Part I: Experimental study of the relation between vortex shedding and transition frequencies',. *Journal of CSME,,* 19(2), 159-167.

[59] Ahmed, N. A. (2006). Investigation of dominant frequencies in the transition Reynolds number range of flow around a circular cylinder Part II: Theoretical determination of the relationship between vortex shedding and transition frequencies at different Reynolds numbers',. *Journal of CSME,,* 19(3), 317-326.

[60] Shun, S., & Ahmed, N. A. (2008). Utilizing wind and solar energy as power sources for a hybrid building ventilation device. Renewable Energy June , 33(6), 1392-1397.

[61] Findanis, N., & Ahmed, N. A. (2008). The interaction of an asymmetrical localised synthetic jet on a side supported sphere. *Journal of Fluids and Structures,* 24(7), 1006-1020.

[62] Longmuir, M., & Ahmed, N. A. (2009). Commercial Aircraft Exterior Cleaning Optimization. *AIAA, Journal of Aircraft,* 46(1), 284-290.

[63] Lien, S. J., & Ahmed, N. A. (2010). Numerical simulation of rooftop ventilator flow. *Building and Environment,* 45, 1808-1815.

[64] Lien, S. J., & Ahmed, N. A. (2011). Effect of inclined roof on the airflow associated with a wind driven turbine ventilator. *Energy and Buildings,* 43, 358-365.

[65] Lien, J., & Ahmed, N. A. (2011). An examination of the suitability of multi-hole pressure probe technique for skin friction measurement in turbulentflow',. in press,, *Journal of Flow Measurement and Instrumentation*, 22, 153-164.

[66] Wu, C., & Ahmed, N. A. (2011). Numerical Study of Transient Aircraft Cabin Flowfield with Unsteady Air Supply. *AIAA Journal of Aircraft*, 48(6), 1994-2002.

[67] Findanis, N., & Ahmed, N. A. (2011). Three-dimensional flow reversal and wake characterisation of a sphere modified with active flow control using synthetic jet. *Advances and Applications in Fluid Mech,,* 9(1), 17-76.

[68] Behfarshad, G., & Ahmed, N. A. (2011). Vortex flow asymmetry of slender Delta Wings',. *International Review of Aerospace Engineering,,* 4(3), 184-188.

[69] Behfarshad, G., & Ahmed, N. A. (2011). Reynolds Stress Measurement Over Four Slender Delta Wings',. *International Review of Aerospace Engineering,,* 4(4), 251-257.

[70] Ahmed, N. A. (2011). Detection of Separation bubble using spectral analysis of fluctuating surface pressure',. *International Review of Aerospace Engineering',,* 4(4).

[71] Behfarshad, G., & Ahmed, N. A. (2011). Effect of unsteady and sinusoidally varying free stream on turbulent boundary layer separation. *Advances and Applications in Fluid Mechanics,* 10(2), 79-98.

[72] Behfarshad, G., & Ahmed, N. A. (2011). Experimental Investigations of Sideslip Effect on Four Slender Delta Wings',. *International Review of Aerospace Engineering,,* 4(4), 189-197.

[73] Ahmed, N. A., & Page, J. R. (2011). Real-time Simulation as a new tool in Future Advanced Aerospace Project Design and Manufacturing Processes. *Advanced Materials Research,* 317-319, 2515-2519.

[74] Ahmed, N. A., & Page, J. R. (2011). Developing and integrated approach to advanced aerospace project design in tertiary education. *Advanced Materials Research,* 317-319, 2520-2529.

[75] Riazi, H., & Ahmed, N. A. (2011). Numerical investigation of four orifice synthetic jet actuators,. *International Review of Aerospace Engineering',,* 4(5), 272-276.

[76] Shun, S., & Ahmed, N. A. (2011). Airfoil Separation Control using Multiple Orifice Air Jet Vortex Generators. *AIAA Journal of Aircraft,* 48(6), 1994-2002.

[77] Ahmed, N. A. (2012). Engineering solutions towards cost effective sustainable environment and living'. *Journal of Energy and Power Engineering,,* 6(2), 155-167.

[78] S. Shun and N.A. Ahmed (2012), 'Rapid Prototyping of Aerodynamics Research Models', Applied Mechanics and Materials vols. 217-219, pp 2016-2025,Trans Tech Publications, Switzerland

[79] Behfarshad, G., & Ahmed, N. A. (2012). Investigation of Newtonian liquid jets impacting on a moving smooth solid surface. *Advances and Applications in Fluid Mechanics*, 12(1).

[80] S. Shun and N.A. Ahmed (2012), 'Design of a Dynamic Stall Test Rig', Applied Mechanics and Materials Vols. 215-216, pp 785-795, Trans Tech Publications, Switzerland

[81] Ahmed, N. A. (2012). Novel developments towards efficient and cost effective wind energy generation and utilization for sustainable environment',. Renewable and Power Quality Journal, 0217-2038X, (10), PL4.

[82] Y.Y.Zheng, N.A.Ahmed and W.Zhang (2012), 'Feasibility Study of Heat Transfer with Fluidic Spike', International Review of Aerospace Engineering, vol. 5, no.2, pp 40-45.

[83] Y.Y.Zheng, N.A.Ahmed and W.Zhang (2012), Impact Analysis of Varying Strength Counter-flow Jet Ejection on a Blunt Shaped Body in A Supersonic Flow, Advances and Applications in Fluid Mechanics, vol 12, no.2, pp 119-129.

[84] Yen, J., & Ahmed, N. A. (2012). Enhancing Vertical Axis Wind Turbine Safety and Performance Using Synthetic Jets',. (in Press),, *Journal of Wind and Industrial Engineering*,.

[85] G. Matsoukas, N. A. Ahmed (2012), 'Investigation of Ionic Wind as a Means of Generating Propulsive Force', International Review of Aerospace Engineering, vol. 5, no. 2, pp 35-39.

[86] Yen, J., & Ahmed, N. A. (2012). Synthetic Jets as a Boundary Vorticity Flux Control Tool', (in press). *AIAA Journal*,.

[87] C.Wu and N.A.Ahmed (2012), 'A Novel Mode of Air Supply for Aircraft Cabin Ventilation', *Building and Environment*, Vol. 56, pp. 47-56

[88] Flynn, T. G., & Ahmed, N. A. (2005). Investigation of Rotating Ventilator using Smoke Flow Visualisation and Hot-wire anemometer',. *Proc. of 5th Pacific Symposium on Flow Visualisation and Image Processing*, [PSFVIP-5], 27-29, September, Whitsundays, Australia, Paper.

[89] Yen, J., & Ahmed, N. A. (2012). Parametric Study of Dynamic Stall Flow Field with Synthetic Jet Actuation. *ASME Journal of Fluids Engineering*, 134, 071106-071101.

Chapter 3

Design Features of a Low Turbulence Return Circuit Subsonic Wind Tunnel Having Interchangeable Test Sections

N. A. Ahmed

Additional information is available at the end of the chapter

1. Introduction

Wind Tunnels have played and are continuing to play a significant role in providing controlled test facilities in the aerodynamic research and development [1-43, 122-178]. The present chapter describes in detail, the design features of a subsonic return circuit wind tunnel that is currently in operation at the Aerodynamics Laboratory of the University of New South Wales. It can be considered to be a general purpose low speed tunnel with a sufficiently large contraction ratio. It has a number of removable turbulence reduction screens to achieve low turbulence level. It also has the provision of removable principal test section and three alternative test section arrangements located at various parts of the wind tunnel circuit. The wind tunnel can provide a wind speed in the range of 0-170 ft/sec at the lowest turbulence level. The top speed can be 200 ft/sec, if a higher turbulence level and spatial non-uniformities produced by omission of the screens can be tolerated.

Floor space limitations of approximately of 65 ft x 12 ft have meant that the tunnel be vertical in the vertical plane. From such consideration and ease of wind tunnel experiments, the test section was placed at the laboratory floor level and the return circuit above the test section. The upper structure of the laboratory roof was too flimsy and inaccessible for satisfactory location of the fan and drive in that area so that the fan and the drive had to be at the floor level. The fan is, therefore, placed downstream of the test section and first diffuser and upstream of the first cascade corner. This unconventional arrangement is not, however, without precedent; similar layout has been used in the N.B.S. 4.5 ft low turbulence wind tunnel and Wichita University 10 ft x 7 ft wind tunnels [44-46].

Figure 1. Side View of the Subsonic Wind Tunnel of the University of New South Wales

2. General considerations

The configuration chosen presents several design advantages as well as disadvantages. These are detailed below:

Advantages:

1. Because the fan is located in a comparatively high speed portion of the tunnel, a favourable flow coefficient for a given tip speed may be more easily obtained, leading to high rotor efficiency

2. Except in the case of high lift or very bluff models, good inlet flow conditions to fan are obtained. This situation does not always occur in tunnels with the conventional fan location immediately after the second cascade corner. Maldistribution of flow may exist due to faulty turning vane performance or the need to pass the fan rotor drive shaft through the second cascade turning vanes. This, in turn, leads to reduced rotor perform-ance and increased noise levels.

3. Flow disturbances created by the fan and its tail fairing in the conventional arrangement may adversely affect the performance of the main return circuit diffuser and hence the wind tunnel. The closed circuit type of diffuser is very sensitive to malfunctions in this diffuser [44,46-48]

4. The long flow return path between the fan and test section aids in achieving a low open tunnel turbulence level. This permits a reduction in the number of screens for certain types of test.

Disadvantages:

1. Since the fan is located in line of sight of the test section, care must be taken in the fan design to keep the noise level at the lowest possible value. Sound waves cause air motions which produce an effect similar to that of turbulence and this may place a lower limit on the tunnel turbulence level [44, 45, 49-51] In a tunnel with conventional fan location, the higher noise frequencies are partly attenuated by the two sets of turning vanes separating the fan and test section. Sound power transmitted to the test section from the fan may, however, be reduced by the tunnel breather slot or the use of ducts with acoustic absorbent inserts [49].

2. Since, for reasons of safety, the fan must be observed in the design of the screen to prevent its causing a high energy loss.

3. Care must be exercised in the design of the fan prerotator blades (if fitted) to render them comparatively insensitive to flow changes caused by the presence of high lifting or bluff models in the test section. The contraction ratio was approximately 7:1, similar to one employed in the N.B.S. 4.5 ft tunnel.

Considerable difficulties had to be overcome in the erection of the tunnel components, none the least of which were the strengthening of the comparatively light floor and roof structures of the laboratory so as to absorb lifting and installation stresses. In its present configuration, the tunnel has an overall length of 67.5 ft, an overall height of 27.5 ft and an overall length of 11.5 ft, excluding interchangeable test sections. Various components of the wind tunnel were built over a period, and the overall work from the start of design to manufacture of various components to final installation took about five years to complete.

3. Design of various components of wind tunnel

The detailed design of the tunnel components is described in the following sections of this report.

3.1. Test section design

The principal test section of 50 inch x 36 inch cross section has the normal value of its width to height, i.e., $\sqrt{2}$:1 [44]. Wall corrections are readily available for this configuration. The test section length of 9.75 ft is within the recommended range for general purpose work of 2.5 to 3 times the equivalent diameter (3.94 ft). Test section fillets, having a side of 5 inches are installed to prevent poor corner flow and accommodate the test section fluorescent lighting.

The test sections are tapered a total of 7/16 inch at the downstream and so as to compensate for the negative static pressure gradient associated with boundary layer thickness increase along the flow. This correction, which was found to be unattainable by tapering the test section fillets, as is sometimes recommended, is calculated to be approximately correct at a test section speed of 160 to 180 m/s. A filtered breather slot is located downstream of the test section.

When the original layout was developed, provisions were made to provide arrangements for removable test sections in various parts of the tunnel circuit. Four such test sections have been provided for. The possible configuration for each of the four is described below:

1. A principal test section having dimensions of cross section of 50 inch x 36 inch and 9.75 ft long and a speed range of 20 to 200 ft/sec.

2. A large test section can be inserted between the screen box/settling chamber assembly and the contraction, the latter being rolled back on a rail system after removal of the principal test section. This large test section is an octagon having maximum dimensions of 10 ft x 10 ft x 9.75 ft and a speed range of from 2 to 30 ft/sec. This test section is useful for a range of industrial aerodynamics tests.

3. An open jet test section, in conjunction with an appropriate removable collector, to be used if required, by removal of the principal test section.

4. A vertical test section which may be interposed in the tunnel circuit in place of the fourth diffuser. This test section permits testing in a vertical airstream and is of octagonal section having maximum dimensions of 5.1 ft x 5.1.ft and a speed range of from 10 to 100 ft/sec.

Of the above four, the first two have been constructed. The test sections were constructed of waterproof quality plywood of either ¾ inch or 1 inch thick, supported on angle from frames. Large viewing windows are provided from ½ inch and ¾ inch thick Perspex set in aluminium frames. The principal test section is provided with doors which open up one complete side over a length of 5 ft and extend two-thirds of the way across the top of the test section to improve accessibility. The tunnel floor is provided with a 3 ft diameter incidence change turntable mounted on a wire bearing race and controlled by a worm and piston drive. The principal test section is removed by means of an overhead travelling trolley and rail system. The large test section is traversed into position by means of a transverse floor rail system which aligns the walls and then by a set of translation tables which move the test section axially forward approximately 4 inches to close the pressure seal. Tapered dowel pins are used to secure accurate alignment of internal airline surfaces and over centre clamps are used to secure the vertical sections together.

3.2. Screen settling chamber design

Wind tunnel screens are required to perform at least two functions, that is, to reduce the:

1. test section turbulence level, and

2. airstream spatial non-uniformities before entrance into the contraction and test section

3.2.1. Turbulence reduction

It has been shown experimentally by Schubauer et al [52] that no turbulence is shed by a screen if the Reynolds number based on the wire material is less than 30 to 60, the exact value depending upon the mesh size and wire diameter. Thus to obtain a low test section turbulence level, the turbulence reduction screens must be placed in a low speed region well upstream of the test section and contraction must consist of wires of the smallest diameter that are consistent with the strength required.

Batchelor [53] reports from experimental work that 'u' and 'v' turbulence components are reduced by factors of 0.36 and 0.54 respectively for wire screens having a resistance co-efficient of 2.0. According to additional experimental work by Dryden and Schubauer [B6), the mean turbulence intensity is reduced by the factor of 0.58 for $k=2.0$ screen and they propose the following relationship based on experiment but confirmed by appropriate theory:

$U'_1/U'_3 = (1+k)^{-0.5}$

U'_1 and U'_3 are the mean turbulence intensities before and after the screens respectively. The relationship between the screen open area ratio or porosity and resistance co-efficient is best found from the data of Annand [54].

The analysis of Batchelor and Drydoen and Schubaureer reveal that it is best to employ a number of screens in series and that of Batchelor indicates that it is the reduction of 'v' component which is most difficult. Relation of the 'v' component to the required level will automatically ensure that the 'u' component is reduced to a correspondingly low value.

3.2.2. Flow non-uniformity reduction

The screens are also required to reduce the flow spatial non-uniformities before the airstream enters the contraction.

A theoretical analysis by Batchelor [55] and an earlier analysis by collar [56] have shown that for steady non-uniform flow, the U component non-uniformities are reduced in the ratio:

$(2-K)/(2+k)$

This expression implies that if $k=2$, the non-uniformities are completely removed. The analysis by Batchelor [53] indicates that the reduction factor can be more accurately expressed as:

$(1-\alpha + \alpha K)/(1+\alpha + K)$

where α is the screen deflection coefficient defined as the ratio of (air exit angle)/(air entry angle)

Taking the approximate value of α [56], the reduction factor for $k=2$ and $\alpha = 0.64$ is seen to be 0.1. Batchelor gives the theoretical reduction for V or transverse velocity non-uniformity component as α or 0.64 for a screen of resistance coefficient 2.0.

3.2.3. Limitations on screen arrangement

Batchelor analysis indicates that the 'v' component of turbulence is reduced by increasing k to a value of 4. However, screens having a resistance coefficient greater than 2 are not normally used, particularly for the final screen, for the following reasons:

1.   Non-uniformity of weave of high resistance coefficient commercial screen materials produce flow disturbances which can have an adverse effect on test section flow distribution and turbulence level

2.   Works by Bradshaw [57], Patel [58] and De Bray [59] have revealed that the final screen open-area ratios of less than 60% are likely to cause the development of flow instabilities of the type described by Morgan [60]. These instabilities produce small angular deviations in the flow downstream of the screens. De Bray suggests that a system of helical vortices originates at the screens and persists through the contraction and interacts with the test section boundary layers. The ultimate effect is to cause lateral variations in thickness and skin friction distribution in the test section boundary layers. Patel also reports that a similar effect is apparent if the screens are allowed to accumulate a build up of dust. Although a single screen resistance coefficient of 2.0 implies screen porosities of about 50%, it is necessary to use, at least for the final screen, a resistance coefficient of approximately 1.4 at 30 ft/sec in order to achieve a porosity of 57%. This is equivalent to a 20 mesh by 30 or 31 gauge wire screen.

There is also evidence to suggest that test section boundary layer disturbance of the type previously mentioned may be avoided by the use of a precision honeycomb located downstream of the last screen [B1,B9,B17]. However, such a device must have very small cell sizes, be of precision, and hence costly, construction and must be located in a very long settling length upstream of the contraction so as to reduce test section turbulence to a value equivalent to that obtained by the use of screen alone.

If screens are used, the attainment of a low turbulence level requires that use of several turbulence reduction screens each with a resistance co-efficient of less than 2. Following suggestion by Perry [B10], it appears reasonable to optimise the screen configuration by the selection of individual screen resistance coefficients which give the maximum reduction in turbulence intensity and spatial non-uniformity with the minimum overall loss. However, in this tunnel, four screens of equal porosity give almost the optimum performance

3.2.4. Screen spacing and settling length

Because of space limitations, it is not usual in wind tunnel design to allow the full length between the turbulence reduction screens required for complete decay of the turbulence introduced by the screen wires. Dryden and Abbott [45] suggest that the turbulence is of the order of the wire diameter wire at a distance of about 200 wire diameters downstream of a screen. A survey of various designs [51] indicates that inter-screen settling lengths to wire diameter ratios of as little as 250 are used. Dryden and Schubauer [62] found that no measurable effect on the test section turbulence level of the N.B.S. 4 ½ ft tunnel was observed when the inter-screen spacing was varied from 2 to 28 inches. Bradshaw and Pankhurst [44] suggest a distance of 500 wire diameters.

The parallel length after the last screen should, however, be as long as possible, consistent with the space available. Most designs for low turbulence wind tunnels appear to have minimum values of about 2000 to 3000 wire diameters [51]. Work of Manton and Luxton [63] shows that the final period of turbulent decay is reached after a distance of approximately 700 wire spacings.

The University of New South Wales 4 ft x 3 ft wind tunnel has a provision for four removable turbulence reduction screens which have an inter-screen settling length of 400 wire diameters and a final settling length of 2000 wire diameters based on the use of 30 gauge wire gauge. A larger final settling length could not be achieved due to inadequate allowance for the screens and turning vanes in the original aerodynamic layout. However, a removable screen facility permits a considerable variety in screen settling length arrangements. The final screen was 20 mesh by 30 or 31 gauge wire and the remaining screens were the same to reduce turbulence and spatial non-uniformities with minimum overall pressure loss.

Because of the long return path between the fan and test section and the closeness of the vane spacing in the fourth cascade, the empty tunnel turbulence level was of the order of 0.2 to 0.3 %, falling to 0.08 to 0.1% with four screens fitted. The similar N.B.S. tunnel had had a turbulence level of 0.26% without screens, decreasing to 0.04% with six screens fitted.

The screen box of the University of New South Wales tunnel is manufactured from ¾ inch waterproof quality plywood reinforced by steel angle iron frames. The wire screens are clamped by bolting between removable pairs of 3 inch x 2 inch Oregon frames which are a neat sliding fit between pairs of similar fixed frames. The movable frames are supported on overhead tracks by sets of small ball-bearing wheels. Ample space has been provided around the edges of the screen box to install spring loaded screen tensioners, or individual frame air seals. The removable frames are provided with adjustable transverse stops and quick acting clamps so as to ensure their accurate and rigid alignment. The screen box door is sealed by a refrigeration type hollow rubber seal and is locked in position by means of eight swing over bolts and large hand wheels. Extensions of the screen sliding tracks are provided outside the screen box to enable the screens to be removed easily.

4. Contraction design

A large contraction ratio is desirable for many reasons, some of which are:

1. A low air speed is obtained in the settling chamber thus permitting the installation of several low loss turbulence reduction screens without excessive power absorption

2. Because of the resulting low air speed in the settling chamber, turbulence generated in the last screen is lower for a given wire diameter

3. For a well designed contraction, the ratio of turbulence intensity to the mean speed will decrease as the mean speed increases at the test section entrance

4. A large contraction ratio, in conjunction with several damping screens, renders the tunnel test section characteristics least susceptible to disturbance in the tunnel circuit, such as those caused by high lift or bluff models [44].

In general, modern wind tunnels are designed for very low turbulence levels require contraction ratios of 12 to 16, in conjunction with up to six turbulence reduction screens. However, quite low turbulence levels may be obtained in wind tunnels with a contraction ratio of the order of 7:1, with four to six screens, and in conjunction with closely spaced vanes in the corner upstream of the settling chamber, as for example, in the N.B.S. 4 ½ ft tunnel [45].

The contraction ratio selected for the University of New South Wales tunnel produces reduction in the percentage longitudinal velocity non-uniformities by a factor of $1/n^2$ or 0.022 [B19] and of the mean RMS turbulence intensity by a factor of the order of [45] :

$U'/U_T = [(2n/3+1/3n^2)^{0.5}]/n = 0.31$

Taylor's alternative analysis suggests 0.4 to 0.8 [53]

There is as yet, no established exact design method for octagonal section wind tunnel contractions. Nevertheless, a design criterion common to all contraction is the avoidance of high wall curvature and large wall slope leading to possible adverse pressure gradients of strength sufficient to cause flow separation in either the contraction or test section.

This problem is particularly critical at the contraction entrance [43 and 46] and modern wind tunnels no longer use very small radius of curvature at the inlet end as was favoured before 1940 [44, 64 and 65]. It has been shown theoretically [66] that in order to obtain a uniform velocity distribution at exit, the velocity increase along the contraction must be monotonic but this condition is incompatible with the need for a finite contraction length. Most methods of design generally fall into of the five following categories:

1. Specification of an arbitrary contraction shape based on experience and/or the demands of the constructional material

2. A contraction shape given by the flow of a uniform stream about an arrangement of sources, sinks or vortex rings.

3. Specification of velocity distribution along the contraction axis leading to a derived contraction shape

4. Conformal transformation techniques

5. Specification of the contraction boundary velocity distribution in the hodograph plane and transformation to the x, r plane so as to derive the contraction shape in axisymmetric or two-dimensional flow.

Details of these methods can be found in References 64 to 84. The method employed for the University of New South Wales tunnel was to sketch in the shape, keeping in mind the demands of the constructional material techniques selected and the requirements for satisfactory performance [44, 46, 48, 64, 66, 69 and 85]. The contraction length was first estimated from the fact that, for contraction ratios of the order of 6 to 10:1, the ratio [51], the length to major inlet dimension, lies within 0.8 to 1.2.

The inlet and exit radii of curvatures are approximately 8 and 11 ft respectively for the University of New South Wales tunnel. The resultant contraction shape is very similar to that derived from an approximate theoretical solution by Cohen and Ritchie [64]. The contraction shape was approximately checked by the application of finite differences applied to the solution of the Laplace equation in radial symmetry [83]. A model was built and satisfactorily tested to confirm further the assumed design shape.

The contraction of the tunnel was manufactured from ¼ in marine ply, mitred and reinforced at the junction of the octagonal sides and built within accurately shaped frames of 3 inch x 2 inch Oregon. The Oregon frames were mounted at 1 ½ ft centres upon a base consisting of three longitudinal bearers of 6 inch x 4 inch Oregon. Flanged wheels and a rail system are mounted under the contraction to enable it to be moved axially along the tunnel centreline between the settling chamber and first diffuser.

5. Diffuser design

As mentioned in section 1, space limitation prevented the fitting of a controlled rapid expansion and the achievement of the optimum contraction ratio of 12 to 16:1. When it is possible to fit such an arrangement, a variety of flow stabilization methods of varying suitability are available for wide angle diffusers [86-94].

Considerable data is also available for the conventional diffuser design [98-104]. Unfortunately, however, little of this information has direct application to the design of three-dimensional octagonal section wind tunnel diffusers of any practical compact design must entail a certain amount of guess work or knowledge of previous experience in the selection of appropriate diffuser angles. For example, attempts to use the data of Ref D6 would indicate that for the large return diffuser of area ratio of 2.85:1, two-dimensional diffuser angles of up to 12^0 might be employed. However, experience with the square cross-section three-dimensional main return diffuser of the R.A.E. No. 2, 11 ½ ft x 8 ½ ft, wind tunnel1 indicated that equivalent cone angles of about 5^0 are satisfactory for this application. Shorter diffusers may employ somewhat larger angles and advantage has been taken of the fact in the design of the University of New South Wales tunnel where the equivalent cone angles used vary from 5.2^0 in the longest diffuser to a maximum of approximately 6 ¼^0 in the shortest diffusers.

The first diffuser downstream of the test section is a particularly difficult design problem as the flow maldistribution caused by high lift and bluff models must be taken into account. Moreover, work by Willis [105] indicates that unsteady flow in the diffuser is responsible for a rise in a measured wall pressure spectra at low frequencies. The University of New South Wales tunnel has an essentially two-dimensional first diffuser with an included angle of 7 ¼ ^{0} and area ratio of 1.4:1 (equivalent cone angle of 3.4^0). Reference D6 indicates that a diffuser angle of up to 17^0 might be employed without separation for this diffuser.

Diffuser performance is also related to the inlet boundary layer thickness and free stream turbulence level [98, 99, 101-104). This makes the estimation of tunnel diffuser losses difficult. In the estimation shown in Table 1, the five diffusers contribute 37% of the tunnel loss, the first diffuser alone being about 14% of the tunnel loss. The design of the diffusion zone over the fan tail-fairing is a special problem and has been conveniently summarised by Russel and Wallis [106].

Diffuser numbers 4 and 5 of the University of New South Wales were built from ¾ inch thick exterior waterproof quality plywood with angle iron and 5 inch x 1 inch timber supporting frames. All sections are octagonal in shape as this permitted short length transitions to be made between the main components of the return circuit and circular fan ducting. The mitred sides of the octagons are constructed of 1/ inch ply mounted on 3 in x 2 in Oregon frames inside the main diffuser shell.

Diffuser No.1, the fan ducting and associated transitions are constructed from 16 gauge mild steel sheet which is reinforced with angle iron frames and rectangular bar steel frames and stringers.

Heavy Perspex windows and fluorescent lighting are fitted to enable easy visualisation of flow performance of the tunnel components. Each leg of the tunnel circuit between the turning vane cascades is provided with one or more quick opening doors for easy access. The doors are sealed with circular, foam rubber cord, formed into shape of an 'O' ring.

6. Turning vane design

It is well known that for abrupt rectangular corners, large aspect ratios and larges ratios of turning radius to inlet width are required to reduce the corner loss [107]. This has led to the post-second world war concept of closely spaced turning vanes to provide low loss, compact, wind tunnel corners.

In the past, it has been common to use thick profile aerofoil turning vanes because these can be designed to give air turning passages of approximately constant area, thus avoiding any expansion and possible flow separation around the passage between adjacent turning vanes. Such turning vanes are efficient in operation, but very difficult and expensive to construct. Winter [108] has shown that these thick vanes may be replaced by thin sheet metal turning vanes with little or no increase e in pressure loss at the corner. According to Winter[108], at a Reynolds number of 1.9×10^6 and for the same spacing to chord ratio (s/c) of 0.25, the thin sheet metal vanes reduced the vane loss to about 50% of that thick profiled turning vanes.

There is very little reliable information in the literature relating to turning vane losses for typical wind tunnel applications. The most extensive information is that reported by Salter [109] who obtained experimental data for both aerofoil profile and sheet metal circular arc turning vanes in the Reynolds number range of 6×10^4 to $1.9 \times 1.9 \times 10^5$. It must be noted that the data presented by Salter does not employ the conventional cascade definition of spacing to chord (s/c) ratio in which the vane spacing is measured normal to the line joining the vane trailing edges. Salter defines a gap to chord ratio based on the distance or gap between the vane trailing edges measured normal to the parallel trailing edge tangents. It would appear that this data has been either misinterpreted or not adequately clarified in most of the subsequent literature [44]. Salter's data has been recalculated according to the conventional cascade definition of s/c ratio

The thin circular arc vanes tested by Salter appear to have a minimum loss co-efficient at an s/c ratio of between 0.3 and 0.4. The difference in the magnitude of the loss co-efficient for the Salter type 2 and 3 vanes could be due to the slightly different camber angles, but it is most likely due to the threefold increase in Reynolds number for the type 3 vanes. The series of tests by Ahmed revealed a considerable variation in loss coefficient with Reynolds number up to a value of about 4×10^5 after which the loss coefficient remained essentially constant. The curves designated Salter 2 and 3 are mean loss coefficients for a cascade corner including losses due to boundary layer and secondary flow effects. Salter also measured the loss coefficient for the potential flow region alone. The greater relative difference can be attributed to the fact that the lesser number of vanes and lower aspect ratio of the type 3 vanes contributes to a larger secondary flow loss. Salter concludes that for 90^0, thin circular arc turning vanes, having 10% straight tangent extensions on the leading and trailing edges, the mean loss coefficient should not exceed 0.1 for Reynolds numbers in excess of 2×10^5. Salter recommends that, to ensure flow stability, the gap chord ratio should be about 0.2 with a vane aspect ratio greater than 3. This gap chord ratio of 0.2 corresponds to an s/c ratio of 0.28 by the conventional cascade definition. Also evident from Salter's results is that the optimum s/c ratio for thick aerofoil profile vanes is in the region of 0.5 to 0.6.

The types of thin sheet metal vanes tested by Silberman[110] have a minimum loss coefficient at an s/c ratio of 0.5 to 0.7 depending upon the vane shape. The curves shown represent the loss coefficients in the potential flow region only. Silberman's results for thick vanes indicate a minima at an s/c value of 0.5.

Since s/c is not the only parameter determining the turning vane design for wind tunnels, a choice must be made of either vane spacing 's' or chord 'c'. This apparent variation possible in this choice is exemplified by the values for the fourth cascade corners of two successful wind tunnels of roughly comparable size and performance, i.e., the R.A.E. 4 ft x 3 ft and N.B.S. 4 ½ ft tunnels. For the R.A.E. tunnel, an s/c ratio of 0.26 was selected using thick profiled turning vanes of 30 inch chord. For the N.B.S. tunnel, the s/c ratio was 0.52 with a chord of 2 7/8 inches, employing thin sheet metal vanes. These two designs represent opposite limits of cascade performance. The R.A.E. vanes appear to have been designed for low loss, whereas those of the N.B.S tunnel were designed for low turbulence. The large chord of the R.A.E. vanes implies high Reynolds numbers and lower loss coefficients. In the N.B.S. tunnel1, the smaller blade spacings selected (approximately 1 ½ inches) resulted in a lower turbulence level measured at the screen location. The 'u' turbulence component of the N.B.S. tunnel1 referred to the settling chamber velocity and, measured in the settling chamber downstream of the fourth cascade, was about 2.3% and about 60% greater than the 'v' or 'w' components. This is a favourable design situation as it is the 'v' and 'w' components which are least reduced by passage through the screens and contraction. In the R.A.E. tunnel, the turbulence level in the comparable location was about 5 % and roughly equal for all three components.

It, therefore, appears that wind tunnel turning vanes can be constructed from thin sheet metal circular arcs, having an s/c ratio in the region of 0.28 to 0.35 and a passage aspect ratio of 6 or more. It appears that vanes for more than 90^0 corners should have a camber angle of 86^0 to 87^0 and that they should be set initially at a positive angle of about 3^0 to 4^0 with trailing edge angle of zero relative to the tunnel centreline at exit. The selection of the value of blade spacing depends upon the application envisioned. Low turbulence tunnels require that small blade spacing be used, for example, a spacing dimension of 2 inch or 3 inch would be unreasonable. Tunnels not requiring a low 'open tunnel' turbulence level might employ spacing dimensions of 12 to 24 inches. Additional compromises to be effected are those of cost and structural integrity. Small vane spacings imply a large number of thin vanes of small chord with a resulting high cost and the possibility of vibration occurring due to relatively low vane natural frequency. Tunnels designed for low corner losses might be designed with a relatively large vane spacing and chord in order to ensure Reynolds numbers in excess of about 4×10^5. Salter suggests that a minimum of 20 turning vanes should be used in low loss corners.

The university of New South Wales tunnel employs s/c ratios of 0.25 and 0.27 for the first and the second cascade corners increasing to 0.31 for the third and fourth corners. Blade spacings vary from 2 to 5 inches and the number of turning vanes from 41 to 33 for the first and fourth cascade corners respectively. The maximum and minimum vane Reynolds numbers at design speed are approximately 5×10^5 and 2.4×10^5 for the first and fourth corners respectively. Turning vane t/c ratios vary between 0.7 to 1.5%.

Because the University of New South Wales wind tunnel cross section is octagonal at all cascade corners and the vane chord is an appreciable dimension, special care had to be taken in the design of the junction between the turning vanes and the octagonal fillet so as to prevent the airstream expanding and subsequently contracting in its passage around the junction zone. The problem was solved by the manufacture of special concave and convex cross sections which were fitted in the cascade corner fillets. The shape of these special corner sections was generated so as to provide a straight line intersection normal to the vane span at the junction of each turning vane and the corresponding corner fillet.

All turning vanes were produced from 10 gauge (1/8 inch) mild steel plate by brake pressing. The turning vanes are set in mild steel plate supporting frames which are reinforced with angle iron.

7. Fan and drive system design

The fan must, by reason of its location downstream of the test section, pose certain design problems as outlined before. These relate to noise level and sensitivity to flow maldistribution caused by high lift or bluff models in the test section.

In general, the design methods of Wallis have been employed [111-113], together with additional experimental data [114-115]. A design utilising 100% pre-rotation has been developed in conjunction with N.P.L. type flow straighteners so as to ensure good efficiency over a wide range of flows together with reduced possibility of stall of the cascade corner vanes immediately downstream of the fan nacelle fairing.

The location of the fan in a relatively high speed portion of the tunnel is associated with a mean rotor blade flow co-efficient of 0.56, which approaches the optimum range of flow coefficients for high fan rotor efficiency with the amount of pre-rotation employed. However, there are conflicting fan duty requirements due to the need for relatively high pressure rise and low fan noise level.

As may be calculated from the estimated tunnel pressure loss characteristic, the fan duty required is 3.8 in w.g. pressure rise at a flow of 1f 122,000 CFM. The tunnel coefficient utilisation is:

(test section energy)/ (Σ circuit losses) = 1.6 to 2.3

depending on the number of screens used.

These requirements have led to the selection of an 8-bladed fan rotor of 5 ft diameter, limited to a maximum tip speed of 315 ft/sec. The rotor blade chords vary from 9.9 inches at the root to 6.4 inches at the tip.

The noise spectrum from an axial flow fan can be described as consisting of two components: 'broad band' noise and 'discrete frequency' noise.

Broad band noise is attributed to two basic mechanisms: vortex shedding from blade boundary layers and interactions between the blading and random turbulence in the intake flow. The theoretical analysis of Refs 116 and 117 show that, for rotor blades operating at their design point, the vortex shedding component of broad band noise is proportional to blade relative velocity to the power 5.6 and that the intake turbulence interaction component is proportional to relative velocity to the power of 4. Reduction in broad band noise can thus be realized mainly by keeping flow velocities adjacent to solid boundaries and, specifically, blade tip velocities, to minimum values consistent with satisfactory aerodynamic performance.

Discrete frequency noise is caused by periodic aerodynamic interaction between fixed and moving blade rows. Like broad band noise, discrete frequency noise has two basic mechanisms. These are the force fluctuations on individual blades which arise from variations in mean velocity of the incoming flow. The data in Refs 116-118 indicate that interaction noise is strongly dependent upon pre-rotator-rotor axial spacing and the shape and size of the individual pre-rotator vanes. The axial spacing affects mainly the potential pressure field interaction mechanism and the vane shape, the mean velocity variation mechanism. As an example, the discussion to Ref 118 indicates that the pressure variation due to the wake persistence is still about 10% of the maximum theoretically possible at a distance equal to one stator chord downstream, for typical accelerating cascades. Experimental data seems to indicate that, consistent with satisfactory aerodynamics, interaction noise is considerably reduced by using separations between stator an rotor of three-quarters to one vane chord in conjunction with small vane areas and slender profiles.

Blade sections chosen for the pre-rotators and rotor are C4 compressor sections on circular arc camber lines [112,114-115 and 119]. These sections give high isolated aerofoil lift coefficients at angles of incidence of 3^0 to 4^0 and have a high stalling incidence. The straightener design is based on the use of the symmetrical NACA 0012 section which starts to stall at about $\pm 14^0$ in the isolated aerofoil condition. Pre-rotator blades of cambered plates were considered [117] because of their comparatively low cost but were abandoned in view of their relatively poor performance under off-design conditions when compared with C4 sections.

Another parameter requiring careful selection was the choice of boss ratio as this affects the overall efficiency of the fan and tail fairing diffuser assembly. Due to the proximity of the first cascade corner, this ratio was fixed at a value of 0.4 which is less than optimum for the rotor alone.

The fan rotor blades have been stressed for centrifugal loading, torsional loads and loads due to non-coincident profile centroids and estimates have been made of the blade natural frequencies [120-121]. The fan rotor was dynamically balanced to an effective centre of gravity displacement of 3 to 5 microns.

The fan design requires a power output of 90 HP at 1200 RPM and a variety of fan drive schemes were considered. Thus a 90 HP compound wound DC motor and ancillaries that included switchgear and speed variation equipment were purchased. The Ward Leonard type speed control system proposed presented considerable difficulty in providing tunnel automatic dynamic head control. In addition, aerodynamic problems were encountered in designing the drive arrangement. A conventional shaft drive through the first cascade was at first envisaged but abandoned when it was realised that the required fairing through the cascade turning vanes caused severe blockage of a component which was already heavily loaded aerodynamically. A direct mechanical drive through a right angle bevel gearbox was next considered. However, a large fairing was needed for the drive shaft and problems were encountered in a gearbox design due to high power transmission requirements in a confined space. Alternative drive systems such as eddy-current variable speed couplings and Thyristor controlled DC drives were also investigated. All these units were costly and suffered from the same basic disadvantage that the prime movers, being large, had to be located outside the tunnel and required some sort of drive shaft arrangement through the tunnel structure to the fan rotor.

Thus the feasibility of using a hydraulic drive system was studied. This system comprises an axial piston hydraulic pump driving similar motor unit and is of the same order of cost as the other systems. The system has many advantages, the main ones being:

1. The drive motor is only 10 inches in diameter and 20 inches long for maximum power output of 125 HP at 1200 RPM. It fits radially inside the fan nacelle fairing where the local diameter is 23 to 24 inches. This eliminates aerodynamic problems associated with a drive shaft through the tunnel structure.

2. Automatic tunnel dynamic head control can be obtained with conventional pneumatic control equipment to a repeatability of ± 0.4 %.

3. The motor speed is fully variable from 0 to 1400 RPM by means of a diaphragm actuator and conventional pressure regulator.

4. The hydraulic pump can be driven by a standard 415 volt, 3-phase induction motor, for which installed electrical capacity was available.

The system finally selected consists of a 150 HP induction motor of 92% efficiency, driving a 'Lucas' PM 3000 series, seven axial piston hydraulic pump fitted with servo-control of the swashplate angle. The servo is operated by a standard 3-15 psi diaphragm actuator. The pump provides high pressure oil at approximately 2300 psi which is supplied to, and returned from the motor by 1 ½ inch outside diameter high pressure tubes through the fan straightener and supporting vanes. Oil flow is approximately 3500 GHP and the overall efficiency of the combined pump and rotor unit is of the order of 82 to 85 %, over the complete speed range. The system also includes ancillary equipment such as a 70 gallon oil reservoir, an oil cooler, boost pump and oil filtration equipment. The main disadvantage of the arrangement is high noise level from the rotor. Provision was, therefore, included in the design for reducing noise transmission of both hydraulic pump and motor.

The fan and drive system and first cascade corner are mechanically isolated from the rest of the tunnel structure, and the laboratory floor, so as to prevent the possibility of any vibrations being transmitted to the test section or instrumentation.

The fan rotor is mounted on an overhung bearing assembly supported off the front of the straightener vane assembly. The straightener vanes are manufactured from ¼ inch mild steel plate with radial and longitudinal plate stiffeners which both provide torsional rigidity and define the aerodynamic profile of the straighteners. The front and the rear of the straightener vanes are attached to heavy steel diaphragm plates at the hub. The front diaphragm plate supports a rigid bearing assembly which carries the overhung fan rotor. The rear diaphragm plate carries another diaphragm plate to which is bolted the hydraulic motor. A flexible coupling connects the very short fan rotor drive shaft and the motor output shaft between the front and rear diaphragm plates. Provision is made in the rotor bearing design to absorb the 400 lb rotor thrust loading. The five straightener vanes have bolted-on cast aluminium nose and tail pieces with the sides sheathed in 16 gauge aluminium sheet.

The fan rotor is of built up construction with blades being held in split root fixings which are in turn clamped between mild steel shroud plates. The rotor blades are high quality aluminium alloy castings with large cylindrical root attachments which enable the blades to be adjusted to any angle by releasing the rotor shroud plate clamping bolts.

The pre-rotator vanes are aluminium alloy castings and are clamped between the shroud plates at the roots to form a rigid prerotator drum assembly. The nacelle nose and tail fairings are spun from 16 gauge aluminium alloy sheet. The nose fairing is bolted on to the front of the pre-rotator drum and the tail fairing to the rear diaphragm plate carrying the hydraulic motor.

Estimations have been made of the tunnel air temperature rise due to power dissipation around the circuit. It was found that without any form of tunnel air exchange or heat exchanger, the air temperature rose by as much as 10 to 15^0 C above ambient after a period of operation of about 10 minutes at a speed of 150 ft/s in the principal test section. This may be doubled for long periods of operation at 200 ft/sec.

The tunnel control system is reasonably straight forward. Instrumentation comprises an optical tachometer, electric drive motor anemometer and pressure gauges for hydraulic system. Electrical interlocks are provided against loss of hydraulic boost pressure and inadvertent starting of the hydraulic system with the hydraulic motor set at the maximum speed condition. Possible fan blade failures are provided for by a fan vibration cut-out switch.

8. Safety net design

For safe operation, a wind tunnel fan must have a suitable safety net located immediately upstream of it to prevent models, or tools, passing through the fan blades. The location of the fan in the University of new South Wales tunnel requires that the safety net be located in the relatively high speed portion of the tunnel circuit. This in turn, requires that considerable care is exercised in the aerodynamic design of the safety net.

It is not unusual to find the safety net located before the first cascade corner even in tunnels with conventional fan layout. It is also known that such safety nets can result in considerable tunnel power expenditure. It was found during experiments on the pressure losses in the ARL 9 ft x 7 ft tunnel that the safety screen which was located before the first corner, contributed 28 % to the total losses. This was the largest of any component. However, the safety net used in this case was relatively coarse, interlocked and 'cylcone' wire mesh.

The University of New South Wales tunnel safety screen is conical in shape and inclined at 45^0 to the free stream direction in order to reduce the velocity component normal to the screen. This configuration also ensures that any object stopped by the screen will be forced to the outside against the tunnel walls. The screen is constructed specially from fine gauge stainless steel wire so as to ensure a low pressure loss. One end of the screen is rigidly held whilst the other end is supported on an energy absorbing spring support.

9. Conclusion

A general purpose return circuit low speed wind tunnel has been designed for the Aerodynamics Laboratory of the University of New South Wales. A contraction ratio of 7:1 and four turbulence reduction screens are used. Low turbulence level is achieved with the assistance of some innovative design features. The fan is located upstream of the first corner. Corner cascade and screen configurations have received special attention.

Other unusual aspects of the design are three sizes of interchangeable test sections in the speed ranges of 0-25 ft/sec, 0-100 ft/sec and 0 -200 ft /sec.

The fan is driven by a hydraulic motor which considerably simplifies power transmission and control problems in this application.

Acknowledgements

The Author wishes to gratefully acknowledge the hard works and dedication of Barry Motson and the late Associate Professor Archer in the Design of this Wind Tunnel

Author details

N. A. Ahmed

School Of Mechanical and Manufacturing Engineering, University of New South Wales, Sydney, NSW, Australia

References

[1] Findanis, N, & Ahmed, N. A. Wind tunnel 'concept of proof' investigations in the development of novel fluid mechanical methodologies and devices', invited Chapter, in 'Wind Tunnels and Experimental Fluid Dynamics Research', edited by J.C. Lerner and U.Boldes, published by In-Tech Organization, Austria, 978-9-53307-623-2July, (2011).

[2] Ahmed, N. A. Wind driven Natural-Solar/Electric Hybrid Ventilators', in 'Wind Power', Section D: The Environmental Issues", Chapter 21, edited by S.M.Muyeen Kitami, published by In-Tech Organization, Austria, 978-9-53761-981-7February, (2010).

[3] Lien, J, & Ahmed, N. A. Wind driven ventilation for enhanced indoor air quality', invited Chapter, in 'Chemistry, Emission, Control, Radiaoactive Pollution and Indoor Air Quality', edited by Nicholas A Mazzeo, published by In-Tech Organization, Austria, 978-9-53307-570-9June, (2011).

[4] Ahmed, N. A, Elder, R. L, Foster, C. P, Jones, J. D. C, & Novel, A. D Laser Anemometer for Boundary Layer Studies', ASME Conf, Boston, USA, 15th December (1987). Also in the 3rd International Symposium on Laser Anemometry, edited by A Dybs et al, ASME, The Fluids Engineering Division, , 55, 175-117.

[5] Ahmed, N. A, Elder, R. L, Foster, C. P, & Jones, J. D. C. Miniature Laser Anemometer for 3D Measurements', J of Measurement Sc Technol, (1990). , 1, 272-276.

[6] Ahmed, N. A, Elder, R. L, Foster, C. P, & Jones, J. D. C. Miniature Laser Anemometer for 3D Measurements', Engineering Optics, (1990). , 3(2), 191-196.

[7] Ahmed, N. A, Elder, R. L, Foster, C. P, & Jones, J. D. C. Laser Anemometry in Turbomachines', IMechE Proc, Part G, J of Aerospace Engineering, (1991). , 205, 1-12.

[8] Ahmed, N. A, Hamid, S, Elder, R. L, Foster, C. P, Jones, J. D. C, & Tatum, R. Fibre Optic Laser Anemometry for Turbo machinery Applications', Optics and Lasers in Engineering, nos 2 and 3, (1992). , 15, 193-205.

[9] Ahmed, N. A, & Elder, R. L. Flow Behaviour in a High Speed Centrifugal Impeller Passage under Design and Off-design Operating Conditions', Fluids and Thermal Engineering, JSME International series B, February, (2000). , 43(1), 22-28.

[10] Simpson, R. G, Ahmed, N. A, & Archer, R. D. Improvement of a Wing Performance using Coanda Tip Jets', AIAA Journal of Aircraft, (2000). , 37(1), 183-184.

[11] Gatto, A, Ahmed, N. A, & Archer, R. D. Investigation of the Upstream End Effect of the Flow Characteristics of a Yawed Circular Cylinder',The RAeS Aeronautical Journal, March, (2000). pp125-128, 104(1033), 253-256.

[12] Gatto, A, Ahmed, N. A, & Archer, R. D. Surface Roughness and Free stream Turbulence Effects on the Surface Pressure over a Yawed Circular Cylinder', AIAA Journal of Aircraft, September, (2000). , 38(9), 1765-1767.

[13] Ahmed, N. A, & Archer, R. D. Performance Improvement of a Bi-plane with Endplates', AIAA Journal of Aircraft, March-April, (2001). , 38(2), 398-400.

[14] Gatto, A, Byrne, K. P, Ahmed, N. A, & Archer, R. D. Pressure Measurements over a Cylinder in Crossflow using Plastic Tubing', Experiments in Fluids, (2001). , 30(1), 43-46.

[15] Ahmed, N. A, & Archer, R. D. Post-Stall Behaviour of A Wing under Externally Imposed Sound', AIAA Journal of Aircraft, September-October, (2001). , 38(5), 961-963.

[16] Ahmed, N. A, & Archer, R. D. Testing of a Highly Loaded Horizonatal Axis Wind Turbines designed for Optimum Performance', International Journal of Renewable Energy, January, (2002). , 25(4), 613-618.

[17] Simpson, R. G, Ahmed, N. A, & Archer, R. D. Near Field Study of Vortex Attenuation using Wing Tip Blowing', The Aeronautical Journal, March, (2002). , 102

[18] Ahmed, N. A, & Goonaratne, J. Lift augmentation of a low aspect ratio thick wing at a very low angle of incidence operating in ground effect", AIAA Journal of Aircraft, March-April (2002). , 39(2)

[19] Ahmed, N. A. Implementation of a momentum integral technique for total drag measurement', International Journal of Mechanical Engineering and Education, (2002). , 30(4)

[20] Pissasale, A, & Ahmed, N. A. Theoretical calibration of a five hole probe for highly three dimensional flow', International Journal of Measurement Science and Technology, July, (2002). , 13, 1100-1107.

[21] Pissasale, A, & Ahmed, N. A. A novel method of extending the calibration range of five hole probe for highly three dimensional flows", Journal of Flow Measurement and Instrumentation, March-April, (2002). , 13(1-2), 23-30.

[22] Ahmed, N. A, & Wagner, D. J. Vortex shedding and transition frequencies associated with flow around a circular cylinder", AIAA Journal, March, (2003). , 41(3), 542-544.

[23] Rashid, D. H, Ahmed, N. A, & Archer, R. D. Study of aerodynamic forces on rotating wind driven ventilator', Wind Engineering, (2003). , 27(1), 63-72.

[24] Ahmed, N. A. An acoustic energy concept for the design of a flow meter', International Journal of Vibration and Acoustics, March (2003). , 8(1), 52-58.

[25] Pissasale, A, & Ahmed, N. A. Examining the effect of flow reversal on seven-hole probe measurements', AIAA Journal, (2003). , 41(12), 2460-2467.

[26] Pissasale, A, & Ahmed, N. A. Development of a functional relationship between port pressures and flow properties for the calibration and application of multi-hole probes to highly three-dimensional flows', Experiments in Fluids, March, March, (2004). , 36(3), 422-436.

[27] Shun, S, & Ahmed, N. A. Utilizing wind and solar energy as power sources for a hybrid building ventilation device', Renewable Energy, June (2008). , 33(6), 1392-1397.

[28] Findanis, N, & Ahmed, N. A. The interaction of an asymmetrical localised synthetic jet on a side supported sphere', Journal of Fluids and Structures, (2008). , 24(7), 1006-1020.

[29] Longmuir, M, & Ahmed, N. A. Commercial Aircraft Exterior Cleaning Optimization', AIAA, Journal of Aircraft, Jan-Feb issue, (2009). , 46(1), 284-290.

[30] Lien, S. J, & Ahmed, N. A. (2010). Numerical simulation of rooftop ventilator flow. Building and Environment, , 45, 1808-1815.

[31] Lien, S. J, & Ahmed, N. A. (2011). Effect of inclined roof on the airflow associated with a wind driven turbine ventilator. Energy and Buildings, , 43, 358-365.

[32] Lien, J, & Ahmed, N. A. An examination of the suitability of multi-hole pressure probe technique for skin friction measurement in turbulent flow', in press, Journal of Flow Measurement and Instrumentation, (2011). , 22, 153-164.

[33] Wu, C, & Ahmed, N. A. Numerical Study of Transient Aircraft Cabin Flowfield with Unsteady Air Supply, AIAA Journal of Aircraft, Nov-Dec issue, (2011). , 48(6), 1994-2002.

[34] Findanis, N, & Ahmed, N. A. Three-dimensional flow reversal and wake characterisation of a sphere modified with active flow control using synthetic jet', Advances and Applications in Fluid Mech, (2011). , 9(1), 17-76.

[35] Behfarshad, G, & Ahmed, N. A. Effect of unsteady and sinusoidally varying free stream on turbulent boundary layer separation', Advances and Applications in Fluid Mechanics, (2011). , 10(2), 79-98.

[36] Shun, S, & Ahmed, N. A. Airfoil Separation Control using Multiple Orifice Air Jet Vortex Generators', AIAA Journal of Aircraft, Nov-Dec issue, (2011). , 48(6), 1994-2002.

[37] Ahmed, N. A. Engineering solutions towards cost effective sustainable environment and living' Journal of Energy and Power Engineering, February (2012). , 6(2), 155-167.

[38] Behfarshad, G, & Ahmed, N. A. Investigation of Newtonian liquid jets impacting on a moving smooth solid surface', Advances and Applications in Fluid Mechanics, (2012). , 12(1)

[39] Ahmed, N. A. Novel developments towards efficient and cost effective wind energy generation and utilization for sustainable environment', Renewable and Power Quality Journal, 0217-2038X, April issue, (2012). (10), PL4.

[40] Zheng, Y. Y, Ahmed, N. A, & Zhang, W. Impact Analysis of Varying Strength Counter-flow Jet Ejection on a Blunt Shaped Body in A Supersonic Flow, (in press) Advances and Applications in Fluid Mechanics,(2012), 12(2), 119-129

[41] Yen, J, & Ahmed, N. A. Parametric Study of Dynamic Stall Flow Field with Synthetic Jet Actuation', Journal of Fluids Engineering, Transactions of the ASME, (2012), 134 (7), 2012,45-53.

[42] Yen, J, & Ahmed, N. A. Synthetic Jets as a Boundary Vorticity Flux Control Tool', AIAA Journal, (2013), 51(2),510-513

[43] Wu, C, & Ahmed, N. A. A Novel Mode of Air Supply for Aircraft Cabin Ventilation', Building and Environment,(2012), 56, 47-56

[44] Bradshaw, P, & Pankhurst, R. C. The Design of Low Speed Wind Tunnels', NPL Aero Report 1039, (1962).

[45] Dryden, H. L, & Abbott, I. H. The Design of Low Turbulence Wind Tunnels', NACA Report 940, (1949).

[46] Razak, K. The University of Wichita 7 ft x 10 ft Wind Tunnel', University of Wichita Engg Report 022, April, (1950).

[47] Squire, H. B, & Winter, K. G. The RAE 4 ft x 3 ft Experimental Low Turbulence Wind Tunnel, Part I: General Flow Characteristics', ARC R&M 2690, February, (1948).

[48] MacPhailD.C., et el., 'The ft x 8 ½ ft Wind Tunnel at RAE, Farnborough', ARC R & M 2424, August, (1945). (2)

[49] Schun, H. The RAE 4 ft x 3 ft Experimental Low Turbulence Wind Tunnel, Part IV: Further Turbulence Measurements', ARC R&M 3261, June, (1953).

[50] Bradshaw, P. Measurements of Free Stream Turbulence in some Low Speed Tunnels at NPL', R & M 3317, (1962).

[51] Ferris, D. H. Measurements of Free Stream Turbulence in the RAE Bedford 13 ft x 9 ft Wind Tunnel', RAE Aero Report 1066, July, (1963).

[52] Schubauer, G. B. et el 'Aerodynamic Characteristics of Damping Screens', NACA TN (2001).

[53] Bachelor, G. K. Homogenous Turbulence', Cambridge University Press, (1953).

[54] Annand, P. The Resistance to Air Flow of Wire Gauges', J Royal Aero Soc, March, (1953). , 141-146.

[55] Bachelor, G. K. On the Concept and Properties of the Idealised Hydrodynamic Resistance', ACA Report ACA-13, (1945).

[56] Collar, A. R. The Effect of a Gauge on Velocity Distribution in a Uniform Duct', ARC R&M 1867, (1939).

[57] Bradshaw, P. The Effect of Wind Tunnel Screens on 2 D Boundary Layers', NPL Aero Report 1085, December, (1963).

[58] Patel, N. P. The Effects of Wind Tunnel Screens and Honey Combs on the Spanwise Variation and Honeycombs on the Spanwise Variation of Skin Friction in 2D Turbulent Boundary Layers', McGill University Mech Engg Tech Note October, (1964). , 64-7.

[59] De Baray, B. G. Some Investigations into the Spanwise Non-uniformity of nominally 2D Incompressible Boundary Layers Downstream of Gauge Screens', ARC Fluid Memo, FM 3863, ARC 29271, July (1967).

[60] Morgan, P. G. The Stability of Flow Through Porous Screens', J of Roy Aero Soc, June (1960). , 359-362.

[61] Lumley, J. L. Passage of a Turbulent Stream Through Honeycomb of Large Length-to Diameter Ratio', Tran ASME, Series D, June (1964). , 218-220.

[62] Dryden, H. L, & Schubauer, G. B. The Use of Damping Screens for the Reduction of Wind Tunnel Turbulence', J of Aero Sci., April (1947). , 221-228.

[63] Manton, M. J, & Luxton, R. E. Note on the Decay of Isentropic Turbulence', Inst of Engineers Austr. Conf. On Hydraulics and Fluid Mechanics', November, (1968). , 93-97.

[64] Cohen, M. J, & Ritche, N. J. B. Low Speed 3D Contraction Design', J of Roy Aero Soc, April, (1962). , 66, 232-236.

[65] Batchelor, G. K, & Shaw, F. S. A Consideration of the Design of Wind Tunnel Contractions', Aust Council Aeronautics Report, ACA-4, (1944).

[66] Cohen, I. A. An Experimental Comparison of the Flow Induced in the Working Section of a Wind Tunnel by Contractions having 2D and 3D Flow Characteristics', Monash Univ Report, MME/65/1, August, (1965).

[67] Thwaites, B. On the Design of Contractions for Wind Tunnels', ARC R & M 2278, March, (1946).

[68] Cheers, F. Notes on Wind Tunnel Contractions', R & M 2137, March, (1945).

[69] Gibbins, J. C, & Dixon, J. R. D Contracting Duct Flow', Quarterly J of Mech and Applied Maths', (1957). , 10, 24-41.

[70] Stanitz, J. D. Design of 2D Channels with prescribed Velocity Distribution along the Channel Walls', Part I: Relaxation Solutions, NACA TN 2593, 1952, Part II: Solutions by Green's Function, NACA TN 2595, (1952).

[71] Smith, A. M. O, & Pierce, J. Exact Solutions of the Newmann Problem: Calculation of Non-Circulatory Plane and Axially Symmetric Flows about or Within Arbitrary Boundaries', Douglas Aircraft Co. Inc Report ES26988, April (1958).

[72] Gibbins, J. C. Design of an Annular Entry to a Circular Duct', Aero Qtly, (1959). , 10, 361.

[73] Libby, P. A, & Reiss, H. R. M. The Design of 2D Contraction Sections', QAM, vol IX, April (1951).

[74] Jackson, J. D. A Description of some Wind Tunnel Contraction Design Data which has been obtained using the Ferranti Mercury High Speed Digital Computer', ARC Report 23, (1962).

[75] Harrop, R. A Method of Designing Wind Tunnel Contactions', J of Roy Aero Soc., (1951). , 55, 169-180.

[76] Smith, R. H, & Wang, C. T. Contracting Cones giving uniform Throat Speeds', J of Aero Sc, October, (1944). , 11

[77] Jordinson, R. Design of Wind Tunnel Contractions', Aircraft Engg, October, (1944). , 33, 294-297.

[78] Whitehead, L. G, Wu, L. Y, & Waters, M. H. L. Contracting Ducts of Finite Length', Aero Qtly, February (1951). , 2, 254-271.

[79] Tsien, H. S. On the Design of the Contraction Cone for a Wind Tunnel', J of Aero Sci., (1943). , 10, 68-70.

[80] Szczeniowski, B. Contraction Cone for a Wind Tunnel', J of Aero Sci., (1943). , 10, 311-313.

[81] Lighthill, M. J. A new Method of 2D Aerodynamic design', ARC R & M., 2112, (1945).

[82] Lilley, G. M. Some Theoretical Aspects of Nozzle Design', M.Sc. Thesis, UL, (1945).

[83] Woods, L. C. A new Relaxation treatment of Flow with Axial Symmetry', Qtly J of Mech and Appl Maths, Part 3, (1951). , 4, 358-370.

[84] Lau, W. T. F. An Analytical Method for the Design of 2D Contactions', J of Roy Aero Soc, January, (1964). , 68

[85] Salter, C, & Raymer, W. C. The NPL 7 ft Wind Tunnel: Sundry Notes and Comments following Measurements of Flow Distribution, Wall Pressures etc.,' NPL Aero Note 1023, October (1963).

[86] Schubauer, G. B, & Spangenberg, W. G. Effect of Screens in Wide Angle Diffusers', NACA Report 949, (1949).

[87] Squire, H. B, & Hogg, h. Diffuser Resistance Combinations in Relation to Wind Tunnel Design', RAE Report Aero 1933, ARC Report &628, April (1944).

[88] Feir, J. B. The Effects of an Arrangement of Vortex Generators Installed to Eliminate Wind Tunnel Diffuser Separation', UTIAS TN june (1965). (87)

[89] Feil, O. G. Vane Systems for Very Wide Angle Subsonic Diffusers', Trans ASME Series D, J of Basic Engg, December, (1964). , 759-764.

[90] Moore, C. A, & Kline, S. J. Some Effects of Vanes and of Turbulence in 2D Wide Angle Subsonic Diffusers', NACA TN 4080, June (1958).

[91] Cochran, D. L, & Kline, S. J. Use of Short Flat Vanes for Producing Efficient Wide Angle Subsonic Diffusers', NACA TN 4309, June (1958).

[92] Fox, R. W, & Kline, S. J. Flow Regime Data and Design Methods for Curved Subsonic Diffusers', Tans ASME, Series D, (1962). , 84, 302-316.

[93] Kline, S. J, Moore, C. A, & Cochran, D. L. Wide Angle Subsonic Diffusers of High Performance and Diffuser Flow Mechanisms', J of Aero Sci, June (1957). , 24(6), 469-470.

[94] Migay, V. K. Investigations of Finned Diffusers', USAF Systems Command,FTD-TT-October, (1962). , 63-86.

[95] Wollett, R. R. Preliminary Investigation of Short 2D Subsonic Diffusers', NANA RM E56CO2, May, (1956).

[96] Raju, K. N, & Rao, D. M. Experiments on the Use of Screens and Splitters for Flow Control in Wide Angle Diffuser', Indian N.A.L., TN-AE-April (1964). , 24-64.

[97] Rao, D. M, & Raju, K. N. Experiments on the use of Splitters for Control in Wide Angle Diffusers', Indian N.A.L. TN-AE-December (1964). , 26-64.

[98] Henry, J. R, et al. Summary of Subsonic Diffusers', NACA L56F05, October (1956).

[99] Cockrell, D. J, & Markland, E. A Review of Incompressible Diffuser Flow', Aircraft Engg, October, (1963). , 286-292.

[100] Patterson, G. N. Modern Diffuser Design', Aircraft Engg, September, (1938). , 267-273.

[101] Cockrell, D. J, & Markland, E. The Effects of Inlet Conditions on Incompressible Fluid Flow in Conical Diffusers', Int J of Mech Sci, (1966). , 8, 125-139.

[102] Kline, S. J, Abbott, D. E, & Fox, R. W. Optimum Design of Straight walled Diffusers', Tans ASME J of Basic Engg, Paper (1959). (58-A)

[103] Winternitz, F. A. L, & Ramsay, W. J. Effects of Inlet Boundary Layer on Pressure Recovery, Energy Conversion and Losses in Conicl Diffusers', J of Roy Aero Soc, February, (1957). , 61, 116-124.

[104] Waitman, B. A, Reneau, R. L, & Kline, S. J. Effect of Inlet Conditions on Performance of 2D Subsonic Diffusers', Trans ASME J of Basic Engg, Paper (1960). (60-WA)

[105] Wills, J. A. B. Spurious Pressure Fluctuations in Wind Tunnels', NPL Aero Report 1237, July (1967).

[106] Wallis, R. A. Axial Fans', Newnes, London, (1961).

[107] Patterson, G. N. Corner Losses in Ducts', Aircraft Engg', August (1937). , 205-208.

[108] Winter, K. G. Comparative Tests of Thick and Thin Turning Vanes in the RAE 4 ft x 3 ft Wind Tunnel', ARC R & M 2589, August, (1947).

[109] Salter, C. Experiments on Thin Turning Vanes', ARC R& M 2469, October, (1946).

[110] Silberman, E. Importance of Secondary Flow in Guide Vane Bends', Tech Paper Series B, 3rd Mid Western Conf on Fluid Mechanics, Univ of Minnesota, (1953). (14)

[111] Wallis, R. A. A Rationalised Approach to Blade Element Design', Conf on Hydraulics and Fluid Mechanics, Inst of Engineers, Aust, (1968).

[112] Wallis, R. A. Optimisation of Axial Flow Fan Design', Ins of Engineers, Aust, M & C Trans, May (1968).

[113] Russel, B. A, & Wallis, B. A. A note on Annular Diffuser Design', CSIRO Int Rep July (1969). (63)

[114] Ruglen, N. Low Speed Wind Tunnel Tests on a series of C4 section Aerofoils', ARC Note ARL/A 275, July (1966).

[115] Edmunds, H. G, & Bartlett, W. J. Section Data of C4 Aerofoils', Power Jets Memorandum April, (1940). (M1090)

[116] Myles, D. J. et el., '' The Design of Axial Flow Fans, Part II: Blade geometry for the Rotor and the Stator', DSIR, NEL Report 181, April (1965).

[117] Sharland, I. J. Sources of Noise in Axial Flow Fans', J Sound Vibrations, (1964). , 1(3), 302-322.

[118] Sharland, I. J. Intake Noise from Axial Flow Turbochargers and Compressors', Proc Inst of Mech Engineers', (1968). , 182(3), 73-77.

[119] Garter, A. D. S. Blade Profiles for Axial Flow Fans, Pumps' compressore', Proc I Mech E, (1961). , 175(15), 775-806.

[120] Schilhans, M. J. Bending Frequency of a Rotating Cantilever Beam', Trans ASME J Applied Mech and Maths, March (1958). , 28-30.

[121] RoxbeeCox, H., (ed) 'Gas Turbine principles and Practice', Chapter 12: Vibrations', G. Newnes, London, (1955).

[122] N.A.Ahmed, 'Investigation of dominant frequencies in the transition Reynolds number range of flow around a circular cylinder Part I: Experimental study of the relation between vortex shedding and transition frequencies', Journal of CSME, vol.19, No.2, 2006, pp159-167

[123] N.A.Ahmed, 'Investigation of dominant frequencies in the transition Reynolds number range of flow around a circular cylinder Part II: Theoretical determination of the relationship between vortex shedding and transition frequencies at different Reynolds numbers', Journal of CSME, vol.19, No.3, 2006, pp 317-326

[124] G.Behfarshad and N.A.Ahmed, 'Vortex flow asymmetry of slender Delta Wings', International Review of Aerospace Engineering, Vol.4, No.3, 2011, pp 184-188

[125] G.Behfarshad and N.A.Ahmed, 'Reynolds Stress Measurement Over Four Slender Delta Wings', International Review of Aerospace Engineering, Vol.4, No.4, 2011, pp 251-257

[126] N.A.Ahmed, 'Detection of Separation bubble using spectral analysis of fluctuating surface pressure', International review of Aerospace Engineering', International Review of Aerospace Engineering', vol.4, no. 4, June, 2011

[127] G.Behfarshad and N.A.Ahmed, 'Experimental Investigations of Sideslip Effect on Four Slender Delta Wings', International Review of Aerospace Engineering, Vol.4, No.4, 2011, pp 189-197

[128] N.A.Ahmed and J.R.Page, 'Real-time Simulation as a new tool in Future Advanced Aerospace Project Design and Manufacturing Processes', Advanced Materials Research, Vols. 317-319 , 2011, pp 2515-2519

[129] N.A.Ahmed and J.R.Page, 'Developing and integrated approach to advanced aerospace project design in tertiary education', Advanced Materials Research, Vols. 317-319 , 2011. pp 2520-2529

[130] H. Riazi, and N.A. Ahmed, 'Numerical investigation of four orifice synthetic jet actuators, International Review of Aerospace Engineering', Vol.4, No. 5, 2011, pp 272-276

[131] S. Shun and N.A. Ahmed, 'Airfoil Separation Control using Multiple Orifice Air Jet Vortex Generators', AIAA Journal of Aircraft, vol 48, no.6, Nov-Dec issue, 2011, pp 1994-2002

[132] N.A.Ahmed, 'Engineering solutions towards cost effective sustainable environment and living' Journal of Energy and Power Engineering, Vol 6, No.2, February 2012, pp155-167

[133] S. Shun and N.A. Ahmed, 'Design of a Dynamic Stall Test Rig', Applied Mechanics and Materials Vols. 215-216 (2012) pp 785-795, © (2012) Trans Tech Publications, Switzerland

[134] G.Behfarshad and N.A.Ahmed, 'Investigation of Newtonian liquid jets impacting on a moving smooth solid surface', Advances and Applications in Fluid Mechanics, vol. 12, no.1, 2012

[135] S. Shun and N.A. Ahmed, 'Rapid Prototyping of Aerodynamics Research Models', Applied Mechanics and Materials Vols. 217-219 (2012) pp 2016-2025, © (2012) Trans Tech Publications, Switzerland

[136] N.A.Ahmed, 'Novel developments towards efficient and cost effective wind energy generation and utilization for sustainable environment', Renewable and Power Quality Journal, ISSN 2172-038X, No. 10, April issue, pp PL4, 2012

[137] Y.Y.Zheng, N.A.Ahmed and W.Zhang, 'Feasibility Study of Heat Transfer with Fluidic Spike', International Review of Aerospace Engineering, vol. 5, no.2, 2012, pp 40-45.

[138] N.A.Ahmed, 'New Horizons of Applications of the 21st Century Aerodynamic Concepts from Aerospace to Power Generation and Utilization', Procedia Engineering, Elsevier Publications, vol. 49, 2012, pp 338-347

[139] S. Shun and N.A. Ahmed, "Wind Turbine Performance Improvements Using Active Flow Control Techniques", Procedia Engineering, Elsevier Publications, vol. 49, 2012, pp 83-91

[140] Matsoukas, G., and Ahmed, N.A., 'Experimental Investigation of Employing Asymmetrical Electrodes in Propulsion of Vehicles', Procedia Engineering, Elsevier Publications, vol. 49, 2012, pp 247-253

[141] G.Behfarshad and N.A.Ahmed, 'Splash measurement of Newtonian Liquid Jets Impacting on a Moving Solid Surface' International Conference on Mechanical Engineering and Mechatronics, Ottawa, August 16-18, 2012

[142] Y.Y.Zheng, N.A.Ahmed and W.Zhang, 'Heat dissipation using minimum counter flow jet ejection during spacecraft re-entry', Procedia Engineering, Elsevier Publications, vol. 49, 2012, pp 271-279

[143] Wu, C., and Ahmed, N.A., 'Application of Flow Control Technique for Indoor Ventilation' Procedia Engineering, Elsevier Publications, vol. 49, 2012, pp 135-141

[144] Riazi, H., and Ahmed, N.A., 'Effect of the ratio of specific heats on a small scale solar Brayton cycle', Procedia Engineering, Elsevier Publications, vol. 49, 2012, pp 263-270

[145] Yen. J., and Ahmed, N.A., 'Improving the Safety and Performance of Small-Scale Vertical Axis Wind Turbine', Procedia Engineering, Elsevier Publications, vol. 49, 2012, pp 99-106

[146] Wongpanyathaworn, M., and Ahmed, N.A., 'Optimising louver locations to improve indoor thermal comfort based on natural ventilation', Procedia Engineering, Elsevier Publications, vol. 49, 2012, pp 169-178

[147] Findanis, N., and N.A. Ahmed, 'Control and Management of Particulate Emissions using Improved Reverse Pulse-Jet Cleaning Systems', Procedia Engineering, Elsevier Publications, vol. 49, 2012, pp 338-347

[148] T.G.Flynn, G.Behfarshad and N.A. Ahmed, 'Development of a Wind Tunnel Test Facility to Simulate the Effect of Rain on Roof Ventilation Systems and Environmental Measuring Devices', Procedia Engineering, Elsevier Publications, vol. 49, 2012, pp 239-246

[149] J. Lien and N.A. Ahmed, Numerical evaluation of wind driven ventilator for enhanced indoor air quality', Procedia Engineering, Elsevier Publications, vol. 49, 2012, pp 124-134

[150] N.A. Ahmed, 'Diverse Applications of Active Flow Control', Commissioned for publication in 'Progress in Aerospace Sciences', a commissioned paper by invitation of Journal board, expected publication, 2013

[151] J. Lien and N.A. Ahmed, 'Indoor Air Quality Measurement with the Installation of a Rooftop Turbine Ventilator', Journal of Environment Protection', Vol.3, No.11, November 2012, pp1498-1508

[152] J.Yen and N.A.Ahmed, 'Enhancing Vertical Axis Wind Turbine Safety and Performance Using Synthetic Jets', Journal of Wind and Industrial Engineering, vol.114, 2013, pp12-17

[153] I.H. Salmom and N.A.Ahmed, "Delaying Stall by Acoustic Excitation Using a Vibrating Film Wing Surface", AIAA 22nd Applied Conference and Exhibit, 16-19 August 2004, Providence, Rhode Island, AIAA Paper No. 2004-4962

[154] N.A.Ahmed, "Turbulent Boundary Layer Analysis of Flow in a Rotating Radial Passage", 2nd BSME-ASME International Conference on Thermal Engineering, 2-4 January, 2004, Dhaka, pp 325-333

[155] T.G.Flynn and N.A.Ahmed, 'Investigation of Rotating Ventilator using Smoke Flow Visualisation and Hot-wire anemometer', Proc. of 5th Pacific Symposium on Flow Visualisation and Image Processing, 27-29 September, 2005, Whitsundays, Australia, Paper No. PSFVIP-5-214

[156] N.Findanis and N.A.Ahmed, 'Wake study of Flow over a sphere', 25th AIAA Applied Aerodynamics Conference, San Francisco, USA, 8-10 June, 2006, AIAA-2006-3855

[157] S.J. Lien and N.A. Ahmed, 'Skin friction determination in turbulent boundary layers using multi-hole pressure probes'25th AIAA Applied Aerodynamics Conference, San Francisco, USA, 8-10 June, 2006, AIAA-2006-3659

[158] N.Findanis and N.A.Ahmed, 'A Flow Study Over a Sphere with Localised Synthetic Jet', 12th Australian International Aerospace Congress/12th Australian Aeronautical Conference19-22 March 2007, Melbourne, Victoria Australia

[159] S.J. Lien and N.A. Ahmed, 'A novel method for Skin friction determination using multi-hole pressure probes', 12th Australian International Aerospace Congress/12th Australian Aeronautical Conference19-22 March 2007, Melbourne, Victoria Australia

[160] B. Ahn and N.A. Ahmed, 'Internal Acoustic excitation to enhance the airfoil performance at high Reynolds number', 14th International Conference on Sound and Vibration, 9-12 July, 2007, Cairns, Australia

[161] N.A.Ahmed, 'The Study of Spectral Properties of a Separation Bubble on a Circular Cylinder under the effects of free stream)turbulence 4th BSME-ASME International Conference on Thermal Engineering, 27-29 December, 2008, Dhaka

[162] N.Findanis and N.A.Ahmed 'Active Flow Control Over a Bluff Body Utilising Localised Synthetic Jet Technology , 13th Australian International Aerospace Congress/ 13th Australian Aeronautical Conference, 9 - 12 March 2009, at the Melbourne Convention Centre, Melbourne, Australia

[163] J.Lien and N.A.Ahmed 'Prediction of Turbulent Flow Separation with Pressure Gradient , 13th Australian International Aerospace Congress, 13th Australian Aeronautical Conference, 9 - 12 March 2009, at the Convention Centre, Melbourne, Australia

[164] N.A.Ahmed and J.R.Page, 'An Improved Approach for Future Aerospace Design process', 13th Australian International Aerospace Congress/13th Australian Aeronautical Conference, 9 - 12 March 2009, at the Convention Centre, Melbourne, Australia

[165] N.A.Ahmed, 'Engineering solutions towards cost effective sustainable environment and living', International Conference on Mechanical, Industrial and Energy Engineering, 22-24 December, 2010

[166] T.J.Flynn and N.A.Ahmed, 'An Investigation of Pitot Tube and Multi Hole Pressure Probe Performance Using a Wet Weather Wind Tunnel Test Section'14th AIAC, Melbourne, 28 Feb-3rd March, 2011

[167] C.Wu and N.A.Ahmed, 'Using Pulsed Jet of Fresh Air to Control CO_2 Concentration in an Air Cabin', 14th AIAC, Melbourne, 28 Feb-3rd March, 2011

[168] C.Wu and N.A.Ahmed, 'Aircraft cabin flow pattern under unsteady air supply', 29th AIAA Applied Aerodynamics Conference, Honolulu, Hawaii, USA. 27 June 2011

[169] H. Riazi and N.A. Ahmed, 'Numerical investigation on two-orifice synthetic jet actuators of varying orifice spacing and diameters', 29th AIAA Applied Aerodynamics Conference, Honolulu, Hawaii, USA. 27 June 2011

[170] G.Behfarshad and N.A.Ahmed, 'Turbulent Boundary Layer Separation in Adverse Pressure Gradient', Proceedings of the 23rd Canadian Congress of Applied Mechanics, 5-9 June, 2011, pp 557-560

[171] N.A.Ahmed, 'Novel developments towards efficient and cost effective wind energy generation and utilization for sustainable environment', Plenary Paper, International Conference on Renewable Energies and Power Quality ICREPQ'12, Santiago de Compostela, Spain, 28-30 March, 2012

[172] S. Shun and N.A. Ahmed, "Utilizing Vortex Acceleration to Improve the Efficiency of Air Jet Vortex Generators", 6th AIAA flow control conference, New Orleans, 25-28 June 2012

[173] H. Riazi and N.A. Ahmed, 'Efficiency enhancement of a small scale closed solar thermal Brayton cycle by a combined simple organic Rankine cycle', Proceedings of the ASME 2012 International Mechanical Engineering Congress & Exposition, IMCE2012, November 9-15, 2012, Houston, Texas, USA

[174] J. Olsen, J.R.Page and N.A. Ahmed, 'A Hybrid propulsion system for a light trainer aircraft', *15th Australian International Aerospace Congress*, 25-28 February 2013.

[175] Y.Y.Zheng, N.A. Ahmed and W.Zhang, 'A Feasibility Study of Bow Shock Wave Heat Dissipation using Counter-Flow Jet Activation', *15th Australian International Aerospace Congress*, 25-28 February 2013.

[176] S.Shun and N.A. Ahmed, 'Experimental Investigation of the Effects of Fluid Acceleration upon Air Jet Vortex Generator Performance', *15th Australian International Aerospace Congress*, 25-28 February 2013.

[177] C.Wu and N.A. Ahmed, 'Vectoring of Wall bounded Planar Ventilation Jet with Synthetic Jet Actuator', *15th Australian International Aerospace Congress*, 25-28 February 2013.

[178] J.Yen and N.A. Ahmed, "Parametric study of synthetic jet on dynamic stall flow field using Computational Fluid Dynamics", *15th Australian International Aerospace Congress*, 25-28 February 2013.

Chapter 4

Portable Wind Tunnels for Field Testing of Soils and Natural Surfaces

R. Scott Van Pelt and Ted M. Zobeck

Additional information is available at the end of the chapter

1. Introduction

Wind erosion of soils refers to the detachment, transport, and subsequent deposition of sediment or surface soils by wind. This process is sometimes termed aeolian movement and is responsible for the formation and migration of dunes, soil degradation in agricultural areas, and formation of deep loess deposits in areas downwind from major sediment sources. From cross-bedding in ancient sandstones, it has been determined that aeolian movement of soils and sediments has been occurring for eons and is a natural geomorphic process. Wind erosion affects over 500 million ha of land worldwide and is responsible for emitting between 500 and 5000 Tg of fugitive dust into the atmosphere annually [1]. These fugitive dust emissions contain a disproportional amount of soil organic carbon and plant nutrients and the winnowing and loss of these materials degrades the soil [2, 3].

Much of what we know about aeolian processes comes from wind tunnel-based investigations. The seminal work of Ralph Bagnold was largely conducted in a stationary suction-type wind tunnel 9 m in length [4]. Wind tunnels allow control over the wind and surface factors controlling aeolian movement and thus much more definitive investigations can be conducted in a shorter period of time than in the natural environment where these factors are highly variable in time and space. Other early aeolian researchers used wind tunnels to assess the erodibility of soil surfaces without plant residues based on the texture of the soil and relative abundance of aggregates too large to be entrained by the wind [5]. Large stationary wind tunnels have allowed sufficiently detailed understanding of the physical processes of aeolian movement that predictive models such as the Wind Erosion Equation [6] and the Wind Erosion Prediction System [7, 8] have been developed.

Stationary wind tunnels continue to be used for aeolian research at scales from single grain movement [9] through soil surface scale [10] to landscape scale [11]. The ability to control the humidity of the atmosphere has enabled scientists to study such sensitive processes as the electrostatic interactions between particles and electrical fields generated during aeolian activities [12]. Stationary wind tunnels have also been used to study abrasion effects of wind-driven sands on building materials [13], crop plants [14], bare crusted soil surfaces [15], and soil surfaces with microphytic crusts [16] as well as to compare and calibrate instrumentation for aeolian filed studies [17, 18].

Fugitive dust is perhaps the most visible product of aeolian activity and stationary wind tunnels have been used to study fugitive dust emissions from eroding soils. From wind tunnel testing of crusted soils and aggregates, it has been determined that sandblasting of these otherwise non-erodible features is responsible for much of the dust generated during aeolian events [19, 20]. Soluble salts such as $CaCO_3$ effects on dust emissions have also been investigated in stationary wind tunnels [21] as have complex and vegetated surfaces [22] and specific soils from Death Valley, a major dust source area in North America [23]. Although stationary wind tunnels have great utility, they are limited to testing disturbed soil surfaces that have been removed from their natural setting. The development of field portable wind tunnels has greatly expanded our ability to investigate aeolian processes in the field under controlled conditions.

2. Portable wind tunnels

Over the last six decades, portable wind tunnels have been developed and used on natural soil surfaces to measure the effects of soil surface characteristics and protective cover on soil erodibility and dust emissions [24]. In their simplest form, portable field wind tunnels must have at least three components: 1.) a self contained or at least portable power source such as an internal combustion engine, 2.) a fan or blower to induce air movement and create an artificial wind, and 3.) a working section that trains the wind from the blower over a finite area of soil surface. Portable wind tunnels in which the fan or blower pushes air through the working section are called pusher-type tunnels and if the fan or blower pulls the air through the working section they are called suction-type wind tunnels. Other components may include transition sections between the blower and the working section including a flow conditioning section and instrumentation to measure the wind speed in the working section and/or to capture sediment at the mouth of the working section. A typical portable field wind tunnel is presented in Figure 1.

The use of portable field wind tunnels has been traced back as far as the early 1940s, but the designers and builders did not publish retrievable documentation of their efforts. Austin Zingg, a mechanical engineer with the US Department of Agriculture, was the first to document the design and construction of a portable wind tunnel [25]. This wind tunnel was used to test the erodibility of crop field surfaces [26] and to assess the effects of roughness and drag based on pressure differentials across the soil surface tested [27]. Other early researchers built a portable wind tunnel to test the susceptibility of field-grown crops to abrasion from saltating particles [28]. A small suction-type tunnel was successfully used to test the threshold wind velocity necessary for particle movement on natural surfaces compared with disturbed surfaces and sieved soil [29]. Another very small suction-type portable wind tunnel has been used in Australia to determine the relative dust emission rates for a range of iron ores and road surfaces [30].

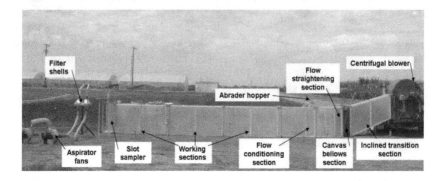

Figure 1. A Typical portable field wind tunnel showing component parts and sampling devices

Australians have also built a truck-mounted portable wind tunnel, tested rectangular and triangular working sections, and determined that the rectangular cross section was superior to the triangular one [31]. These same researchers noted the importance of wind flow conditioning upstream of the working section. Their wind tunnel has been used to assess the erodibility of bare cultivated and uncultivated soil [32], the effects of disturbance on the erodibility of cryptogamic crusts [33], and the sandblast injury and subsequent growth of narrow-leaf lupine [34].

In North America, a pusher-type wind tunnel was built to test the effects of oriented and random surface roughness elements on soil erodibility [35, 36]. This wind tunnel needed a small tractor and a secondary transmission for its power source and was transported using a large truck and 16 m long trailer. Another large portable wind tunnel built in North America was a suction-type wind tunnel that had a 12 m long working section. This wind tunnel was used to determine the erodibilities of natural crusted surfaces in North America and Africa [37-40]. A pusher-type wind tunnel with the power source and blower mounted on a truck bed and the working section lifted from the truck bed and lowered into place on the soil surface by hydraulic arms has been successfully employed to assess dust emissions from loess soils with and without surface cover in the Pacific Northwest of North America [24, 41-44] (Figure 2).

Figure 2. A large wind tunnel working section being lowered into place by a hydraulic arm.

Although large portable wind tunnels requiring mechanical devices to install may be powerful and allow testing of relatively large surface areas, the logistics of transporting them and finding a suitable footprint of level ground to test limit their utility. Examples of medium-size tunnels that may be installed by human power include a German tunnel that was field calibrated [45], a portable boundary layer wind tunnel with a working section formed of three 2 m long elements that fits on a 5 m trailer [46], and another German design that incorporates a rainfall simulator to induce wind-driven rain splash [47]. A summary of portable field wind tunnels, the dimensions of their working sections, maximum wind velocities developed, and reported boundary layer depths is presented in Table 1.

Publication	Tunnel Design	Width (m)	Height (m)	Length (m)	U_{max} (m s⁻¹)	Bdy. Lyr. (m)
Zing [25]	Pusher	0.91	0.91	9.12	17	0.23
Armbrust and Box [28]	Pusher	0.91	1.22	7.32	18	----
Gillette [29]	Suction	0.15	0.15	3.01	7	----
Fryrear [34, 35]	Pusher	0.60	0.90	7.00	20	0.15
Nickling and Gillies [37]	Suction	1.00	0.75	11.90	15	>0.2
Raupach and Leys [31]	Pusher	1.20	0.90	4.20	14	0.40
Pietersma et al. [24]	Pusher	1.00	1.20	5.60	"/>20	>1.0
Leys et al. [30]	Suction	0.05	0.10	1.00	19	----
Maurer et al. [45]	Suction	0.60	0.70	9.40	15	----
Van Pelt et al. [46]	Pusher	0.50	1.00	6.00	19	0.50
Fister and Ries [47]	Pusher	0.70	0.70	3.00	8	0.2

Table 1. Summary of previous and present portable wind tunnel designs, dimensions, maximum wind speed reported, and boundary layer thickness.

3. Wind tunnel design

In an engineering paper on wind tunnel design [48], the author stated that "the design of blower-driven air tunnel…is a combination of art, science, and common sense, the last being the most essential. It is difficult and unwise to predict firm rules for tunnel design." In addition to the power source, fan or blower, flow conditioning section, and working section, flow tripping fences and spires are often used to deepen the boundary layer thickness [49, 50], abrader feeders and regulators are used to initiate a saltation cloud, and sediment samplers to quantify the rate of erosion and dust emissions are often included in the design. Instrumentation such as anemometers of many designs and particle impact sensors are often used to monitor wind tunnel performance and to set operating parameters. In almost all cases, portable field wind tunnel designs are somewhat unique and highly influenced by their intended use.

3.1. Practical design criteria

When Zingg published the design and operation of his first portable field wind tunnel [25], he offered seven practical criteria to consider. These practical criteria are listed below;

1. The wind tunnel must be capable of producing an air stream free of general rotation and of known and steady characteristics.

2. It must provide easy and positive control of a range of wind velocities and forces common to the natural wind.

3. It must be durable.

4. It must be safe to use.

5. It should have sufficient size to afford free movement and representative sampling of eroding materials over field surfaces.

6. It must have ready portability.

7. It should be light in weight and amenable to quick and positive assemblage and dismantling.

Another criterion that he used but did not list was the use of commercially available equipment when available.

3.2. Aerodynamic design criteria

Mike Raupach and John Leys [31] suggested six aerodynamic criteria that should be considered in addition to the seven practical criteria proposed by Zingg. These aerodynamic criteria are listed below:

1. The flow must reproduce the logarithmic wind speed profile in the natural atmosphere, thus ensuring realistic aerodynamic forces on saltating grains.

2. The surface shear stress must scale correctly with the wind speed above the surface so that realistic aerodynamic forces act on grains of all sizes at the surface.

3. The vertical turbulence intensity and scale in the region close to the ground must be realistic, ensuring that vertical turbulent dispersion of suspended grains is properly modeled.

4. The flow must be spatially uniform to avoid local scouring by anomalous regions of high surface stress.

5. Gusts should be simulated in the tunnel due to the fact that higher shear stress is required to initiate erosion than to sustain it.

6. A portable wind tunnel simulation of erosion should allow for the introduction of saltating grains at the beginning of the working section if more than the very upwind area of an eroding field is to be simulated.

They noted that criteria 1 to 4 are satisfied if the air flow near the ground surface is a well developed equilibrium boundary layer sufficiently deep to contain particle motion in the inner region where the mean wind speed profile is logarithmic and uniform over the eroding area. The logarithmic wind speed profile for neutral atmospheric stability has been described by:

$$U_z = \left(u^*/k\right) \ln\left(z/z_0\right) \tag{1}$$

where U is the wind speed at height z above the surface, u^* is the friction velocity, $z_{.}$ is the aerodynamic roughness length of the underlying surface, and k is the von Karman constant, usually assigned a value of approximately 0.4.

Criterion 5 requires turbulence with length scales greater than possible within the practical dimensions of portable wind tunnels and cannot be naturally generated by shear forces within either the working sections or flow conditioning sections of a portable wind tunnel. They tried to simulate gustiness using mechanical interruption of air flow in the flow conditioning section of their tunnel but discovered that the turning vane they employed for this purpose reduced the mean wind speed without increasing the vertical turbulence.

3.3. Simulating saltation

Although criterion 6 is not truly aerodynamic, it is very necessary in order to simulate well developed steady state saltation of sand grains over an eroding surface. However, it also raises more questions as to the design and operation of the portable wind tunnel such as how much material to introduce, what the size distribution should be, and how to distribute it realistically in the flow before it strikes the ground surface tested in the working section. An orifice controlled gravity fed saltation initiator that drops the sand abrader into inclined tubes for acceleration before striking a sandpaper surface and bouncing into the flow stream is shown in Figure 3.

Figure 3. A complex flow conditioning section showing the abrader hopper and inclined tubes used to initiate saltation into the flow stream.

Saltation has been shown to reach a maximum at about 7 m length in wind tunnels [51] and decreases at longer distances, reaching equilibrium at between 10 and 15 m into the working section [45]. Longer working sections have limited utility however due to their lower transportability [47] and require a substantially longer uniform level surface on which to be set

[52]. Working section lengths of portable field wind tunnels have varied from 3 m [19, 47] to almost 12 m [38, 39]. Recently, a small circular device named the Portable In-Situ Wind Erosion Research Laboratory (PI-SWERL) [52] has been used to develop shear stress over a surface and entrain particles using radially induced rather than linearly induced shear stress.

3.4. Power sources

Power sources have ranged from external sources such as the power take-off shaft of a tractor as input to a transmission that output to drive chains [35, 36], to self-contained direct drive internal combustion engines [24, 25, 28, 31, 38-40], self contained internal combustion engines driving hydraulic pumps to provide for a hydraulic drive motor at the blower [46], and electric motors supplied by portable generators [45, 47]. All these power sources are field tested and reliable. The wind speed may be adjusted by varying the engine or motor speed or by changing the pitch of the fan or blower blades.

3.5. Fans and blowers

The fans and blowers employed for wind tunnels are of two primary types. Axial fans (Figure 4a) are composed of fixed or adjustable pitched blades arranged radially around the axis of rotation, which is often aligned with the axis of flow through the wind tunnel. Although axial fans are highly efficient at inducing flow, the flow tends to spiral and this problem must be addressed [53] if the flow conditions of Zingg's first criterion are to be met. Centrifugal blowers (Figure 4b) have fixed pitch blades or impellers that are arranged parallel to the axis of rotation at the circumference of a blower cage. The axis rotation is commonly normal to the axis of air flow down the wind tunnel. Centrifugal blowers tend to be more flexible with respect to design, are more stable and efficient over a variety of flows, and produce less spiraling in the flow than axial fans [53].

Some portable field wind tunnels are too compact for adequate flow conditioning. This shortcoming is very problematic as flow considerations are the most important factor in the successful operation of the wind tunnel [31]. Wind tunnels may not reach true transport capacity or overshoot true transport capacity if flow conditioning upwind of the working section is inadequate [54] and wind tunnel height may limit the amount of upward mixing during strong turbulent diffusion [23]. The height of the working section affects the depth of the boundary layer that may be achieved. Upper limits of the Froude number F have been proposed for wind tunnel design of from 10 [55] to 20 [24]. The Froude number is defined by:

$$F = U^2 / gH \tag{2}$$

Where U is the wind tunnel design wind speed, g is the acceleration due to gravity, and H is the wind tunnel height. A well developed boundary layer at least 50 cm thick is recommended to ensure initiation of vertical particle uplift [45]. For this reason, mini-tunnels and micro-tunnels may be too small to allow results that can be scaled up to field scales [56].

Figure 4. An axial fan (a) and a centrifugal blower (b) typical of those used in construction of portable field wind tunnels.

3.6. Flow conditioning

Flow conditioning sections of various designs have been used to straighten the flow and remove or reduce the scale of eddies in the flow, to initiate a logarithmic wind speed profile and turbulence, and to initiate saltating abrader material into the air flow down the wind tunnels. A typical honeycomb flow straightener with 10 mm screen layers used to create an even logarithmic wind speed profile is presented in Figure 5. If the flow is properly conditioned and the height of the wind tunnel is not limiting the depth of the boundary layer may be estimated from the wind speed profile in the wind tunnel working section [46]. Investigators have stated that although boundary layer thickness is a poorly defined concept, it may be estimated as the height at which the wind speed profile attains 99 percent of its maximum value [57]. Finally, the proper regulation of carefully chosen abrader material allows for saltation clouds representing different rates of erosion and surface abrasion although rates consistent with those noted in the field for natural sand movement [58] are commonly used. Portable field wind tunnel may be used to estimate the threshold wind velocity necessary to initiate particle movement using impact sensors [18] or optically based sensors [59]. The technique of using the percentage of seconds in which moving particles are noted [60] is easily employed in a portable wind tunnel if the wind speed can be slowly and evenly increased.

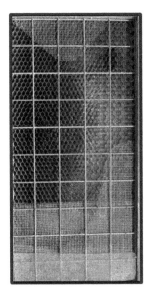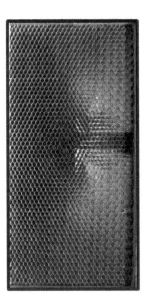

Figure 5. A flow conditioner showing the large cell honeycomb used to break the scale of eddies and straighten flow and also the 10 mm screen layers used to even the flow and create a logarithmic wind speed profile in the wind tunnel.

4. Conclusions

Over the last 6 decades, portable field wind tunnels have been successfully used on several continents to study the controlling processes of aeolian particle movement, assess the erodibility of natural surfaces subjected to different disturbances, estimate dust emission rates for natural surfaces, investigate the partitioning of chemical and microbiological components of the soil on entrained sediment, and to estimate the threshold wind velocity necessary to initiate aeolian particle movement. Although not a perfect replacement for wind in the natural environment due to the absence of turbulent gusts, the forces created by the wind are repeatable and the accuracy of the tunnel is solely dependent on the accuracy of the devices measuring critical operating parameters such as wind velocity and sediment loading. When properly designed, calibrated, constructed, and operated, very useful information can be obtained in a relatively short period of time with these tools.

Acknowledgements

USDA is an equal opportunity provider and employer.

Nomenclature of symbols and their units

U– Wind velocity (m s^{-1})

z– Height above the surface (m)

z$_o$– Roughness length (m)

u* - Friction velocity (m s^{-1})

k– von Karman constant (~0.4)

F– Froude number (dimensionless)

g–Acceleration due to gravity (m s^{-2})

H– Wind tunnel height (m)

Author details

R. Scott Van Pelt[1*] and Ted M. Zobeck[2]

*Address all correspondence to: scott.vanpelt@ars.usda.gov

1 United States Department of Agriculture – Agricultural Research Service (USDA-ARS), Big Spring, Texas, USA

2 USDA-ARS, Lubbock, Texas, USA

References

[1] Grini A, Myhre G, Zender C, Sundet J, Isakssen I. Model Simulations of Dust Source and Transport in the Global Troposphere: Effects of Soil Erodibility and Wind Speed Variability. Institute Report Series No. 124. Norway, University of Oslo, Department of Geosciences 2003.

[2] Zobeck T, Fryrear D. Chemical and Physical Characteristics of Windblown Sediment: II. Chemical Characteristics and Total Soil and Nutrient Discharge. Transactions of the ASAE 1986; 29(4) 1037-1041.

[3] Van Pelt R, Zobeck T. Chemical constituents of fugitive dust. Environmental Monitoring and Assessment 2007; 130 3-16.

[4] Bagnold R. The Physics of Blown Sand and Desert Dunes. London: Methuen; 1941.

[5] Chepil W. Properties of Soil Which Influence Wind Erosion: I. The Governing Principle of Surface Roughness. Soil Science 1950; 69(2) 149-162.

[6]   Woodruff N, Siddoway F. A Wind Erosion Equation. Soil Science Society of America Proceedings 1965; 29(5) 602-608.

[7]   Hagen L. Evaluation of the Wind Erosion Prediction System (WEPS) Erosion Submodel on Cropland Fields. Environmental Modeling and Software 2004; 19(2) 171-176.

[8]   Hagen L, Wagner L, Skidmore E. Analytical Solutions and Sensitivity Analyses for Sediment Transport in WEPS. Transactions of the ASAE 1999; 46(6) 1715-1721.

[9]   Huang N, Zheng X, Zhou Y, Van Pelt, R. Simulation of Wind Blown Sand Movement and Probability Density Function of Liftoff Velocities of Sand Particles. Journal of Geophysical Research 2006; D20201, doi:10.1029/2005JD006559.

[10] Kohake D, Skidmore E, Hagen L. Wind Erodibility of Organic Soils. Soil Science Society of America Journal 2010; 74(1) 250-257.

[11] Offer Z, Goossens D. Wind Tunnel Experiments and Field Measurements of Aeolian Dust Deposition on Conical Hills. Geomorphology 1995; 14(1) 43-56.

[12] Zheng X, Huang N, Zhou Y. Laboratory Measurement of Electrification of Wind-Blown Sands and Simulation of Its Effect on Sand Saltation Movement. Journal of Geophysical Research 2003; 108(D10): 4322 doi:10.1029/2002/D002572.

[13] Liu L, Gao S, Shi P, Li Y, Dong Z. Wind Tunnel Measurements of Adobe Abrasion by Blown Sand: Profile Characteristics in Relation to Wind Velocity and Sand Flux. Journal of Arid Environments 2003; 53(3) 351-363.

[14] Baker J. Cotton Seedling Abrasion and Recovery by Wind-Blown Sand. Agronomy Journal 2007; 99(2) 556-561.

[15] Zobeck T. Abrasion of crusted Soils: Influence of Abrader Flux and Soil Properties. Soil Science Society of America Journal 1991; 55(4) 1091-1097.

[16] McKenna Neumann C, Maxwell C. A Wind Tunnel Study of the Resilience of Three Fungal Crusts to Particle Abrasion During Aeolian Transport. Catena 1999; 38(2) 151-173.

[17] Goossens D, Offer Z. Wind Tunnel and Field Calibration of Six Aeolian Dust Samplers. Atmospheric Environment 2000; 34(7) 1043-1057.

[18] Van Pelt R, Peters P, Visser S. Laboratory Wind Tunnel Testing of Three Commonly Used Saltation Impact Sensors. Aeolian Research 2009; 1(1-2) 55-62.

[19] Gillette D. A Wind Tunnel Simulation of the Erosion of Soil: Effect of Soil Texture, Sandblasting, Wind Speed, and Soil Consolidation on Dust Production. Atmospheric Environment 1978; 12(8) 1735-1743.

[20] Rice M, McEwan I. Crust Strength; A Wind Tunnel Study of the Effect of Impact by Saltating Particles on Cohesive Soil Surfaces. Earth Surface Processes and Landforms 2001; 26(7) 721-733.

[21] Amante-Orozco A, Zobeck T. Clay and Carbonate Effect on Dust Emissions as Generated in a Wind Tunnel. In: Lee J, Zobeck T. (eds.) ICAR5/GCTE-SEN Joint Conference Proceedings 2002, Lubbock, TX, USA.

[22] Kim D, Cho G, White B. A Wind-Tunnel Study of Atmospheric Boundary-Layer Flow over Vegetated Surfaces to Suppress PM_{10} Emissions on Owens (Dry) Lake. Boundary-Layer Meteorology 2000; 97(2) 309-329.

[23] Roney J, White B. Estimating Fugitive Dust Emission Rates Using an Environmental Boundary Layer Wind Tunnel. Atmospheric Environment 2006; 40(40): 7668-7685.

[24] Pietersma D, Stetler L, Saxton K. Design and Aerodynamics of a Portable Wind Tunnel for Soil Erosion and Fugitive Dust Research. Transactions ASAE 1996; 39(6) 2075-2083.

[25] Zingg A. A Portable Wind Tunnel and Dust Collector Developed to Evaluate the Erodibility of Field Surfaces. Agronomy Journal 1951; 43(2) 189-191.

[26] Zingg A. Evaluation of the Erodibility of Field Surfaces with a Portable Wind Tunnel. Soil Science Society of America Proceedings 1951; 15(1) 11-17.

[27] Zingg A, Woodruff N. Calibration of a Portable Wind Tunnel for the Simple Determination of Roughness and Drag on Field Surfaces. Agronomy Journal 1951; 43(2) 191-193.

[28] Armbrust D, Box J. Design and Operation of a Portable Soil-Blowing Wind Tunnel. USDA-ARS Pub. No. 41-131, US Govt. Print Off. Washington, D.C. 1967.

[29] Gillette D. Tests with a Portable Wind Tunnel for Determining Wind Erosion Threshold Velocities. Atmospheric Environment 1978; 12(12) 2309-2313.

[30] Leys J, Strong C, McTainsh G, Heidenreich S, Pitts O, French P. Relative Dust Emission Estimated from a Mini-Wind Tunnel. In: Lee J, Zobeck, T (eds.) ICAR5/GCTE-SEN Joint Conference Proceedings 2002, Lubbock, TX, USA.

[31] Raupach M, Leys J. Aerodynamics of a Portable Wind Erosion tunnel for Measuring Soil Erodibility by Wind. Australian Journal of Soil Research 1990; 28(2) 177-191.

[32] Leys J, Raupach, M. Soil Flux Measurements Using a Portable Wind Erosion Tunnel. Austalian Journal of Soil Research 1991; 29(4) 533-552.

[33] Leys J, Eldridge D. Influence of Cryptogamic Crust Disturbance to Wind Erosion on Sand and Loam Rangeland Soils. Earth Surface Processes and Landforms 1998; 23(11).

[34] Bennell J, Leys J, Cleugh H. Sandblasting Damage of Narrow-Leaf Lupine (Lupinus angustifolius L.): A Wind Tunnel Simulation. Australian Journal of Soil Research 2007; 45(2) 119-128.

[35] Fryrear D. Soil Ridges-Clods and Wind Erosion. Transactions of the ASAE 1984; 27(2) 445-448.

[36] Fryrear D. Soil Cover and Wind Erosion. Transactions of the ASAE 1985; 28(3) 781-784.

[37] Nickling W, Gillies J. Emission of Fine-Grained Particulates from Desert Soils. In: Leinen M, Sarnnthein M. (eds) Paleoclimatology and Paleometeorology: Modern and Past Patterns of Global Atmospheric Transport, Series C: Mathematical and Physical Sciences, 282, Dordrecht, The Netherlands, Kluwer Academic; 1989. p133-165.

[38] Houser C, Nickling W. The Emission and Vertical Flux of Particulate Matter <10 μm from a Disturbed Clay-Crusted Surface. Sedimentology 2001; 48(2) 255-267.

[39] Houser C, Nickling W. The Factors Influencing the Abrasion Efficiency of Saltating Grains on a Clay-Crusted Playa. Earth Surface Processes and Landforms 2001; 26(5) 491-505.

[40] Macpherson T, Nickling W, Gillies J. Dust Emissions from Undisturbed and Disturbed Supply-Limited Desert Surfaces. Journal of Geophysical Research 2008; 113: F02S04, doi:10.1029/2007JF000800.

[41] Saxton K, Chandler D, Stetler L, Lamb B, Claiborne C, Lee B. Wind Erosion and Fugitive Dust Fluxes on Agricultural Lands in the Pacific Northwest. Transactions of the ASAE 2000; 43(3) 623-630.

[42] Chandler D, Saxton K, Busacca A. Predicting Wind Erodibility of Loessial Soils in the Pacific Northwest by Particle Sizing. Arid Land Resource Management 2005; 19(1) 13-27.

[43] Sharratt B. Instrumentation to Quantify Soil and PM_{10} Flux Using a Portable Wind Tunnel. In: Proceedings of the International Symposium on Air Quality and Waste Management for Agriculture. ASABE Paper No. 701P0907cd. St. Joseph, Michigan, USA; ASABE 2007.

[44] Copeland N, Sharratt B, Wu J, Foltz R, Dooley J. A Wood-Strand Material for Wind Erosion Control: Effects on Sediment Loss, PM_{10} Vertical Flux, and PM_{10} Loss. Journal of Environmental Quality 2009; 38(1) 139-148.

[45] Maurer T, Hermann L, Gaiser T, Mounkaila M, Stahr K. A Mobile Wind Tunnel for Wind Erosion Field Measurements. Journal of Arid Environments 2006; 66(2) 257-271.

[46] Van Pelt R, Zobeck T, Baddock M, Cox J. Design, Construction, and Calibration of a Portable Boundary Layer Wind Tunnel for Field Use. Transactions of the ASAE 2010; 53(3) 1413-1422.

[47] Fister W, Iserloh T, Ries J, Schmidt R. A Portable Wind and Rainfall Simulator for In Situ Soil Erosion measurements. Catena 2012; 91(1) 72-84.

[48] Mehta R. The Aerodynamic Design of Blower Tunnels with Wide-Angle Diffusers. Progress in Aerospace Science 1977; 18 59-120.

[49] Counihan J. An Improved Method of Simulating an Atmospheric Boundary Layer in a Wind Tunnel. Atmospheric Environment 1969; 3(2) 197-214.

[50] Irwin H. The Design of Spires for Wind Simulation. Journal of Wind Engineering and Industrial Aerodynamics 1981; 7(3) 361-366.

[51] Shao Y, Raupach M. The Overshoot and Equilibrium of Saltation. Journal of Geophysical Research 1992; 97(D18) 20559-20564.

[52] Sweeney M, Etyemezian V, Macpherson T, Nickling W, Gillies J, Nicolich G, McDonald E. Comparison of PI-SWERL with Dust Emission Measurements from a Straight-Line Wind Tunnel. Journal of Geophysical Research 2008; 113: F01012, doi: 10.1029/2007JF000830.

[53] Mehta R, Bradshaw P. Design Rules for Small Low-Speed Wind Tunnels. Aeronautical Journal 1979; 83 443-449.

[54] Hagen L. Assessment of Wind Erosion Parameters Using Wind Tunnels. In: Stott D, Mohtar R, Steinhardt G (eds.) Sustaining the Global Farm: Selected Papers form the 10th International Soil Conservation Organization Meeting, West Lafayette, IN, USA Purdue University and the USDA-ARS National Soil Erosion Laboratory 2001; p742-746.

[55] White B, Mounla H. An Experimental Study of Froude Number Effect on Wind Tunnel Simulation. In: Barndorff-Nielsen O, Willets B (eds.) Aeolian Grain Transport, Volume I: Mechanics New York, New York, Springer-Verlag 1991; p145-157.

[56] Fister W, Ries J. Wind Erosion in the Central Ebro Basin Under Changing Land Use Management: Field Experiments with a Portable Wind Tunnel. Journal of Arid Environments 2009; 73(11) 996-1004.

[57] Schlichting H, Gertsen K. Boundary Layer Theory. Berlin, Heidelberg, New York, New York, Springer-Verlag 2000.

[58] Namikas S. Field Measurement and Numerical Modeling of Aeolian Mass Flux Distribution on a Sandy Beach. Sedimentology 2003; 50(2) 303-326.

[59] Sherman D, Bailiang L, Farrell E, Ellis J, Cox W, Maia L, Sousa P. Measuring Aeolian Saltation: A Comparison of Sensors. Journal of Coastal Research 2011; 10059 280-290.

[60] Stout J. A Method for Establishing the Critical Threshold for Aeolian Transport in the Field. Earth Surface Processes and Landforms 2004; 29(10) 1195-1207.

Chapter 5

A Method of Evaluating the Presence of Fan-Blade-Rotation Induced Unsteadiness in Wind Tunnel Experiments

Josué Njock Libii

Additional information is available at the end of the chapter

1. Introduction

1.1. Flows driven by a constant pressure gradient through a pipe of circular cross section

When the flow of a Newtonian fluid in a pipe of circular cross section is driven solely by a constant pressure gradient, the resulting velocity distribution is a quadratic function of the radial distance from the axis of the pipe. The velocity profile of such a flow has, therefore, a parabolic distribution in which the maximum velocity occurs on the axis of the pipe. A graphical representation of this type of velocity is shown in Figure 1.

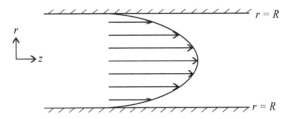

Figure 1. The parabolic velocity profile for flow driven by a constant pressure gradient in a circular pipe

1.2. Flows driven by a sinusoidal pressure gradient through a pipe of circular cross section

Things become more complicated if the pressure gradient varies with time. When, for example, the pressure gradient fluctuates with time in such a way that that gradient can be expressed as a simple sinusoidal function, the velocity profile remains parabolic only at very low frequencies of fluctuation. At very high frequencies, the location of the maximum velocity moves away from the axis of the pipe and towards the wall. The higher the frequency of oscillations of the pressure gradient, the farther away the point of maximum velocity moves from the axis of the pipe. Sample plots of velocity profiles that were generated at high frequencies of fluctuations are shown in the literature by Uchida (1956). Here, Figure 2 is one such example, where five snapshots of velocity profiles at different times are displayed, from left to right, within one complete cycle: at the beginning, one-quarter, half-way, three-quarters of the way, and at the very end of the cycle. The values of the parameters that were used to generate these plots are summarized below:

$$-\frac{1}{\rho K}\frac{\partial p}{\partial x}=cos(nt); \quad k=\sqrt{\frac{n}{v}}R=5; c=\frac{Kk^2}{8n}=3.125\frac{K}{n}$$

Where n is the circular frequency, p the pressure, ρ the mass density of the fluid, t the time, x the axial coordinate, R the inside radius of the pipe, u the axial speed of the fluid, v the coefficient of kinematic viscosity, k a dimensionless ratio used by Schlichting to denote the magnitude of the frequency of oscillation, and K is a constant that indicates the size of the pressure gradient.

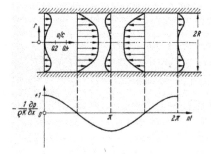

Figure 2. Sample velocity profiles for flow driven by a sinusoidal pressure gradient in a circular pipe [Uchida]

1.3. The mean velocity squared and Richardson's annular effect

The higher the frequency of oscillations of the pressure gradient, the farther away the point of maximum velocity moves from the axis of the pipe. The phenomenon in which the point of maximum velocity moves away from the axis of the pipe and shifts towards its wall is known as Richardson's annular effect. It was demonstrated experimentally by Richardson (1929), proved analytically by Sexl (1930), and demonstrated to hold for any pressure gradient that is periodic with time by Uchida (1956).

When the sinusoidal pressure gradient that drives the flow in a circular pipe has fast oscillations, the mean velocity squared computed with respect to time is found to be

$$\overline{u^2(r)} = \frac{K^2}{2n^2}\left\{1 - 2\sqrt{\frac{R}{r}}\exp\left[-\sqrt{\frac{n}{2v}}(R-r)\right]\cos\left[\sqrt{\frac{n}{2v}}(R-r)\right] + \frac{R}{r}\exp\left[-2\sqrt{\frac{n}{2v}}(R-r)\right]\right\}$$ (1)

Where r is the radial distance from the axis of the pipe; and letting $y = (R - r)$ be a new variable that represents the distance from the wall of that pipe, a dimensionless distance from that wall can be defined as $\eta = y\sqrt{\frac{n}{2v}}$. Using this distance, one can nondimensionalize the mean velocity squared as shown below :

$$\frac{\overline{u^2(r)}}{\left(\frac{K^2}{2n^2}\right)} = \left\{1 - 2\sqrt{\frac{R}{r}}\exp\left[-\sqrt{\frac{n}{2v}}(R-r)\right]\cos\left[\sqrt{\frac{n}{2v}}(R-r)\right] + \frac{R}{r}\exp\left[-2\sqrt{\frac{n}{2v}}(R-r)\right]\right\}$$ (2)

When one is very close to the wall of the pipe, r and R are very close in magnitude and $\frac{R}{r} \approx 1$. This causes the expression in Eq. (2) to become

$$\frac{\overline{u^2(y)}}{\left(\frac{K^2}{2n^2}\right)} = 1 - 2\exp(-\eta)\cos\eta + \exp(-2\eta).$$ (3)

When the variation of the expression of the mean velocity squared in Eq. (3) is plotted against the dimensionless distance η, as shown in Figure 3, one can see that the location of the maximum velocity is not on the axis of the pipe as is the case in steady flow and at very low oscillations of the pressure gradient. Instead, it occurs near the wall of the pipe at a dimensionless distance $\eta = y\sqrt{\frac{n}{2v}} = 2.28$. This value has been shown to agree with experimental data collected by Richardson (1929).

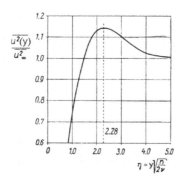

Figure 3. Variation of the mean with respect to time of the velocity squared for periodic pipe flows that are very fast

In this Figure 3, y is the distance from the wall of the pipe and $u_\infty^2 = \frac{K^2}{2n^2}$ represents the mean with respect to time of the velocity squared at a large distance from the wall.

2. Richardson's annular effect in a wind tunnel

Unsteady pulsating flows occur in many situations that have a practical engineering importance. These include high- speed pulsating flows in reciprocating piston-driven flows, rotor blade aerodynamics and turbomachinery. They also arise in wind-tunnel flows. When the velocity distribution is measured across the test section of a subsonic wind tunnel that is driven by a high speed fan, it has been observed experimentally that, in addition to the effect of the boundary layer that is expected near the wall, Richardson's annular effect can be demonstrated as well. Indeed, published experimental results from our laboratory have demonstrated that Richardson's annular effect can occur in a wind tunnel (Njock Libii, 2011).

The purpose of the remainder of this chapter is to summarize the theoretical basis of the Richardson's annular effect in pipes of circular sections and in rectangular tubes, illustrate its results graphically, and relate them to what happens in a wind tunnel.

First Stokes' second problem is reviewed briefly. The theory of pulsating flows in pipes and ducts is summarized. The anatomy of the shift in the location of the maximum velocity from the center to points near the wall is presented using series approximations and graphical illustrations.

3. Stokes' second problem

Fundamental studies of fully-developed and periodic pipe and duct flows with pressure gradients that vary sinusoidally have been done (Sexl, 1930). From such studies, we know that, when an incompressible and viscous fluid is forced to move under a pulsating pressure difference in a pipe or a duct, some characteristic features are always observed. Some of these features are similar to those that are observed to occur in the boundary layer adjacent to a body that is performing reciprocating harmonic oscillations. These features are related to the results of a classic problem solved by Stokes, known as Stokes' second problem, which gives details of the behavior of the boundary layer in a viscous fluid of kinematic viscosity, v, that is bounded by an infinite plane surface that moves back and forth in its own plane with a simple harmonic oscillation that has a circular frequency, ω.

Stokes solution shows that, for this type of flow, 1) transverse waves propagate away from the oscillating surface and into the viscous fluid; 2) the direction of the velocity of these waves is perpendicular to the direction of propagation; 3) the oscillating fluid layer so generated has a phase lag, φ, with respect to the motion of the wall; and 4) that phase lag, which varies with y, the distance from the wall, is given by, $\varphi = \frac{y}{\delta}$, where δ represents a length scale introduced by Stokes; that length, called the depth of penetration of the wave into the fluid, is given by $\delta = \left(\frac{2v}{\omega}\right)^{1/2}$. The thickness of the boundary layer is proportional to the penetration depth, δ. The value of the constant of proportionality varies with the point that one designates to be the edge of the boundary layer. Thus, For example, if one defines the edge of the boundary layer to be the point in the flow where the speed inside the boundary layer become equal to 99% of the speed of flow outside the boundary layer, the constant of proportionality is 4.6. Then, the thickness of the boundary layer at that point is equal to 4.6 δ.

4. Pulsating flow through pipes

4.1. Basic equations

The flow of a viscous fluid in a straight pipe of circular cross-section due to a periodic pressure gradient was examined experimentally and theoretically by Richardson and Tyler (1929) and theoretically by Sexl (1930). If the pipe is sufficiently long, variations of flow parameters along its axis may be neglected and the only component of flow is that along the axis of the pipe. The Navier-Stokes equations become

$$\frac{\partial u}{\partial t} = - \frac{1}{\rho} \frac{\partial p}{\partial x} + \nu \left(\frac{\partial^2 u}{\partial r^2} + \frac{1}{r} \frac{\partial u}{\partial r} \right) \tag{4}$$

$u(r=a, \ t)=0 \ ; and \ u(r=0, \ t)= finite$

$$- \frac{1}{\rho} \frac{\partial p}{\partial x} = a \ function \ of \ time \tag{5}$$

Where u = u(r, t) is the component of velocity in the axial direction x, $\frac{\partial p}{\partial x}$ is the pressure gradient in the axial direction, t is the time, v is the kinematic viscosity of the fluid, r is the radial distance measured from the axis of the pipe, and a is the inside radius of the pipe. For a given pressure gradient, one seeks solutions that are finite at r = 0 and satisfy the no-slip condition u = 0 on the wall of the pipe at all times. We present two cases: First, the case of a sinusoidal pressure gradient that was first solved by Sexl (1930) and then that of a general periodic pressure gradient that was first solved by Uchida (1956).

4.2. Case of a sinusoidal pressure gradient: Sexl's method (1930)

If the pressure gradient is sinusoidal and given the form

$$\frac{\partial p}{\partial x} = \rho C cos(\omega t), \tag{6}$$

then, the solution is given by the real part of

$$u(r, t) = -i\frac{C}{\omega}\left\{1 - \frac{J_o\left((-ix)^{\frac{1}{2}}\frac{r}{a}\right)}{J_o\left((-ix)^{\frac{1}{2}}\right)}\right\}e^{i\omega t} \tag{7}$$

Where J_o is the Bessel function of the first kind and of zero order (Watson,1944) and, here, x is defined as shown below:

$$x = \frac{\omega a^2}{v}. \tag{8}$$

For small values of the parameter x, the real part of the velocity u can be written as

$$u(r, t) = \frac{C}{4v}\left(a^2 - r^2\right)\cos(\omega t) \tag{9}$$

and for large values of the parameter x and of $\left(\frac{r}{a}\right)^2$, the velocity can be represented by

$$u(r, t) = \frac{C}{\omega}\left\{\sin(\omega t) - \left(\frac{a}{r}\right)^{1/2}\exp(-\alpha)\sin[\omega t - \alpha]\right\}; \tag{10}$$

where

$$\alpha = \left(\frac{x}{2}\right)^{1/2}\left(1 - \frac{r}{a}\right). \tag{11}$$

Furthermore, the mean velocity squared computed with respect to time is found to be

$$\overline{u^2(r)} = \frac{C^2}{2\omega^2}\left\{1 - 2\left(\frac{a}{r}\right)^{\frac{1}{2}}\exp(-\alpha)\cos(\alpha) + \left(\frac{a}{r}\right)\exp(-2\alpha)\right\}. \tag{12}$$

These well-known results indicate that the representation of the velocity changes radically as one varies the parameter x from very small to very large values. For example, the maximum velocity reaches its maximum amplitude on the axis of the pipe when x is very small. However, when the frequency of fluctuations becomes large, the location of the maximum velocity shifts away from the axis of the pipe and moves closer and closer to the wall of the pipe as the parameter increases, Fig. 4. Indeed, in the latter case, the expression for the location of maximum velocity is given by

$$r = a\left(1 - 3.22x^{-1/2}\right). \tag{13}$$

4.3. Case of a general periodic pressure gradient: Uchida's general theory

The case of a general periodic pressure gradient was solved by Uchida (1956), whose solution is summarized below.

Consider a general periodic function that can be expressed using a Fourier series as follows:

$$-\frac{1}{\varrho}\frac{\partial p}{\partial x} = \varkappa_0 + \sum_{n=1}^{\infty} \varkappa_{cn} \cos nt + \sum_{n=1}^{\infty} \varkappa_{sn} \sin nt , \tag{14}$$

Where n is the frequency of oscillation and $\varkappa_{cn}$ and $\varkappa_{sn}$ are Fourier coefficients.

In complex form, the solution to Eq. (4) is given by

$$u = \frac{\varkappa_0}{4v}\left(a^2 - r^2\right) - \sum_{n=1}^{\infty} \frac{i\varkappa_n}{n}\left[1 - \frac{I_o\left(kri^{\frac{3}{2}}\right)}{I_o\left(kai^{\frac{3}{2}}\right)}\right]e^{int} \tag{15}$$

Where

$$k = \sqrt{\frac{n}{v}} \tag{16}$$

The total mean velocity U is defined as $U = \frac{G}{\pi a^2}$, where G, the total mean mass flow, is given by

$$G = \frac{1}{2\pi}\int_0^{2\pi}dt\int_0^a 2\pi u r dr = \frac{\pi a^4 \varkappa_0}{8v} . \tag{17}$$

When this expression has been rearranged in order to introduce the mean pressure gradient, one gets

$$G = \frac{1}{2\pi}\int_0^{2\pi}dt\int_0^a 2\pi u r dr = \frac{\pi a^4 \varkappa_0}{8\mu}\overline{\left(-\frac{\partial p}{\partial x}\right)}, \tag{18}$$

Where $\overline{\left(-\frac{\partial p}{\partial x}\right)} = \rho\varkappa_0$, is the mean pressure gradient taken over time. Therefore,

$$U = \frac{a^2}{8\mu}\overline{\left(-\frac{\partial p}{\partial x}\right)}. \tag{19}$$

If one uses U as a velocity scale, the nondimensional expression of the velocity is given by

$$\frac{u}{U} = \frac{u_s}{U} + \frac{u'}{U}$$

(20)

with

$$\frac{u_s}{U} = 2\left(1 - \frac{r^2}{a^2}\right)$$

(21)

And

$$\frac{u'}{U} = \sum_{n=1}^{\infty} \frac{\chi_{cn}}{\chi_0}\left\{\frac{8B}{(ka)^2}\cos nt + \frac{8(1-A)}{(ka)^2}\sin nt\right\} + \sum_{n=1}^{\infty} \frac{\chi_{sn}}{\chi_0}\left\{\frac{8B}{(ka)^2}\sin nt - \frac{8(1-A)}{(ka)^2}\cos nt\right\} ,$$

where

$$A = \frac{ber\,(ka)ber\,(kr) + bei\,(ka)\ bei\,(kr)}{ber^2(ka) + bei^2(kr)} , \quad B = \frac{bei\,(ka)ber\,(kr) - ber\,(ka)\ bei\,(kr)}{ber^2(ka) + bei^2(kr)} \quad \text{(a)}$$

And

$$J_0\left(kri^{\frac{3}{2}}\right) = ber(kr) + ibei(kr) \text{ (b)}$$

(22)

In which *ber* and *bei* are Kelvin functions defined using infinite series as shown below:

$$ber(z) = \sum_{k=0}^{\infty} \frac{(-1)^k\left(\frac{z}{2}\right)^{4k}}{((2k)!)^2} \text{ (c)}$$

and

$$bei(z) = \sum_{k=0}^{\infty} \frac{(-1)^k\left(\frac{z}{2}\right)^{4k+2}}{((2k+1)!)^2} . \quad \text{(d)}$$

4.4. Asymptotic expressions of the velocity distribution

Two extreme cases were considered by Uchida: the case of very slow pulsations and that of very fast pulsations, depending on the magnitude of the dimensionless parameter $ka = \sqrt{\frac{n}{v}}a$

Consider very slow pulsations of the pressure gradients. If $= \sqrt{\frac{n}{v}}a \ll 1$, pulsations of the pressure gradients are very slow. Then, under these conditions and from the behavior of Kelvin functions, it is reasonable to expect that

berka $\rightarrow 1$ *and beika* $\rightarrow 0$.

Then, the velocity takes the form

$$\frac{u}{U} = 2\left(1 - \frac{r^2}{a^2}\right)\frac{1}{\chi_0}\left[-\frac{1}{\rho}\frac{\partial p}{\partial x}\right] = \frac{1}{4\mu}\left(a^2 - r^2\right)\left[-\frac{1}{\rho}\frac{\partial p}{\partial x}\right].$$

(23)

In this case, the velocity distribution is a quadratic function of the radial distance from the axis of the pipe ; and the corresponding velocity profile is parabolic. This result is similar to what is obtained in steady flow. However, the magnitude of the velocity is a periodic function of time and is always in phase with the driving pressure gradient.

Consider very fast pulsations of the pressure gradients. If $ka = \sqrt{\frac{n}{v}}\, a \rightarrow \infty$, pulsations of the pressure gradients are very fast. Then, Uchida used asymptotic expansions of ber(ka) and bei(ka). In this extreme, the expression for the velocity near the center of the pipe is different from that near the wall of the pipe. So, they are discussed separately.

Near the center of the pipe, $ka \rightarrow \infty$ and $kr \rightarrow 0$, one gets

$$\frac{u}{U} = \frac{x_0}{4v}\left(a^2 - r^2\right) + \sum_{n=1}^{\infty}\frac{x_{cn}}{n}\cos\left(nt - \frac{\pi}{2}\right) + \sum_{n=1}^{\infty}\frac{x_{sn}}{n}\sin\left(nt - \frac{\pi}{2}\right). \tag{24}$$

Comparing this to Eq. (14), one sees that when the pulsations are very rapid, fluid near the axis of the pipe moves with a phase lag of 90° relative the driving pressure gradient and its amplitude decreases as the frequency of pulsation increases.

Near the wall of the pipe, $kr \rightarrow ka \rightarrow \infty$, and one uses asymptotic expansions of Bessel functions to get

$$\frac{u}{U} = 2\left(1 - \frac{r^2}{a^2}\right) + \sum_{n=1}^{\infty}\frac{x_{cn}}{x_0}\frac{8}{(ka)^2}\left\{\sin(nt) - \sqrt{\frac{a}{r}}\exp\left(-\frac{k(a-r)}{\sqrt{2}}\right)\sin\left[nt - \frac{k(a-r)}{\sqrt{2}}\right]\right\}$$
$$+ \sum_{n=1}^{\infty}\frac{x_{sn}}{x_0}\frac{8}{(ka)^2}\left\{-\cos(nt) + \sqrt{\frac{a}{r}}\exp\left(-\frac{k(a-r)}{\sqrt{2}}\right)\cos\left[nt - \frac{k(a-r)}{\sqrt{2}}\right]\right\}. \tag{25}$$

4.5. Case of a general periodic pressure gradient: Graphical illustrations of Uchida's results

Uchida presented graphical illustrations of these results for four different values of the parameter ka: 1, 3, 5, and 10.

At each value of the parameter ka and using the angle, nt, as the variable, he plotted twelve different snapshots of the velocity profiles of the unsteady component of velocity for the following angles:

$nt = 0^0,\ 30^0,\ 60^0,\ 90^0,\ 120^0,\ 150^0,\ 180^0,\ 210^0,\ 240^0,\ 270^0,\ 300^0,\ 330^0.$

His plots showed that, as the value of ka was increased, the location of maximum velocities shifted progressively away from the axis of the pipe and moved towards the wall. At ka = 1, all maximums of velocity distributions occurred on the axis of the pipe. At ka = 3, two maximums of velocity distributions had shifted away from the axis and moved toward the wall of the pipe. These occurred at nt = 0° and nt = 180°. At ka = 5, half the maximums of velocity distributions had shifted away from the axis and moved toward the wall of the pipe. These occurred at nt = 0°, 30°, 60°, 180°, 210° and 240°. At ka = 10, all of the maximums of velocity

distributions had shifted away from the axis and occurred by the wall of the pipe. These results are summarized in Table 1 and Uchida's (1956) plots are reproduced in enlarged for-
mats in Figures 5(a), 5(b), 6(a), and 6(b), as shown below.

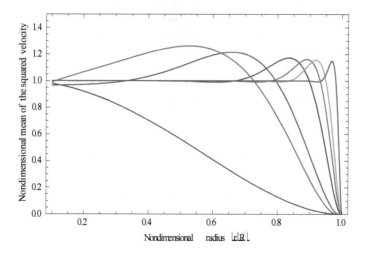

Figure 4. As the frequency of pressure pulsations increases, the point of maximum velocity shifts progressively away
from the axis of the pipe and moves towards its wall (plots of Eq. (2), for increasing values of n).

ka	Total maximums	Maximums on the axis of the pipe	Maximums away from the axis of the pipe
1	12	12	0
3	12	10	2
5	12	6	6
10	12	0	12

Table 1. Data extracted from Uchida's papers (his Figures 1, 2, 3, and 4 are shown below).

5. Pulsating flow through rectangular ducts

5.1. Summary of the results of analysis

Yakhot, Arad, and Ben-dor conducted numerical studies of pulsating flows in very long rectangular ducts, where a and h were the horizontal and the vertical dimensions, respectively,
of the cross-section of the duct, Fig. 7. Letting $\alpha = \left(\frac{\omega}{2v}\right)^{1/2}$, they performed calculations for
low and high frequency regimes ($1 \leq \alpha h \leq 20$) in rectangular ducts using two different as-

pect ratios (a/h =1 and a/h = 10). They presented results for low frequencies (ah =1) and

moderate frequencies (ah =8). They indicated that results for frequencies higher, $ah \geq 10$,

were very similar to those for moderate frequencies. The other conclusions that they came

up with are summarized below.

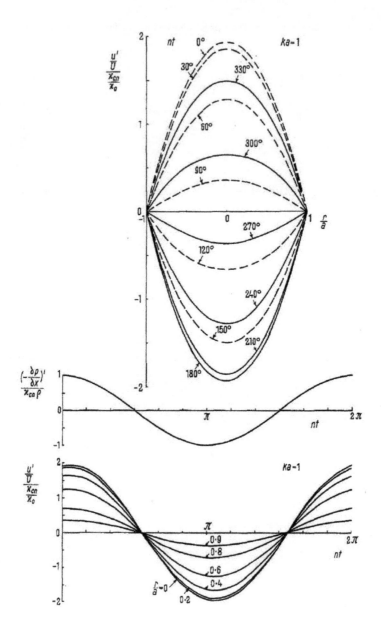

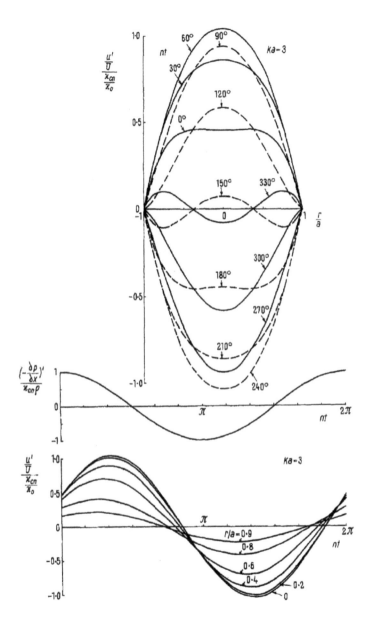

Figure 5. (a). Where maximums of velocity distributions occur when the parameter ka = 1. The angle nt is the parameter; in these plots, $nt = 0°$, $30°$, $60°$, $90°$, $120°$, $150°$, $180°$, $210°$, $240°$, $270°$, $300°$, $330°$. (b). Where maximums of velocity distributions occur when the parameter ka = 3. The angle nt is the parameter; in these plots, $nt = 0°$, $30°$, $60°$, $90°$, $120°$, $150°$, $180°$, $210°$, $240°$, $270°$, $300°$, $330°$.

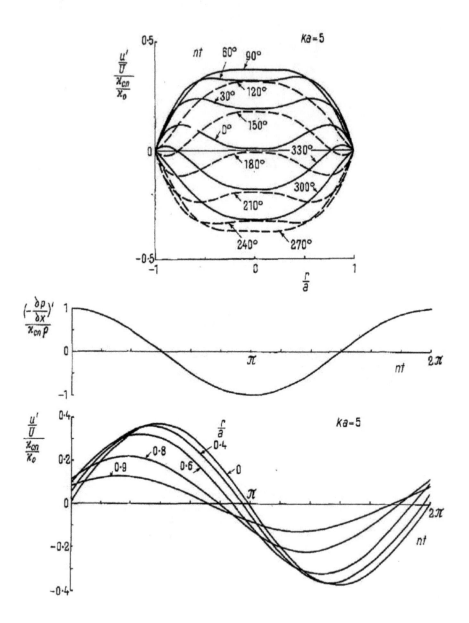

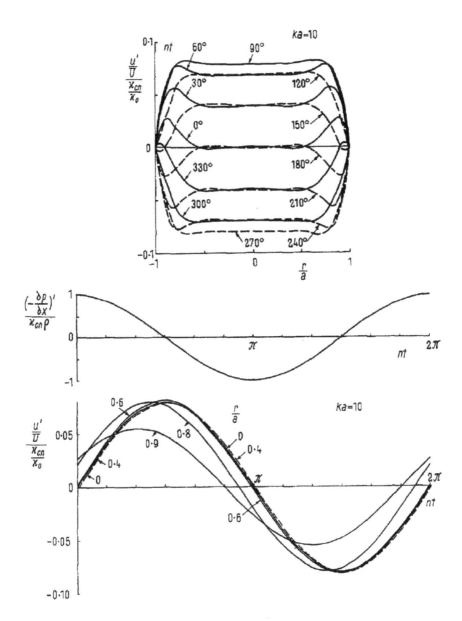

Figure 6. (a) Where maximums of velocity distributions occur when the parameter ka = 5. The angle nt is the parameter; *in these plots*, nt = 0°, 30°, 60°, 90°, 120°, 150°, 180°, 210°, 240°, 270°, 300°, 330°. (b). Where maximums of velocity distributions occur when the parameter ka = 10. The angle nt is the parameter; *in these plots*, nt = 0°, 30°, 60°, 90°, 120°, 150°, 180°, 210°, 240°, 270°, 300°, 330°.

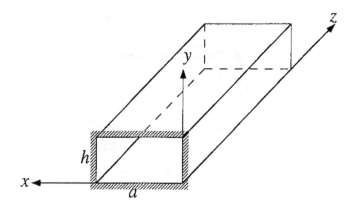

Figure 7. Sketch of the rectangular duct used by Yakhot, Arad and Ben-dor (1999) in their numerical studies.

For low pulsating frequencies, $h = 1$, flow in a duct of square cross-sectional area, the velocity distribution is in phase, that is in lock step, with the driving pressure gradient.
This was true at low and at high aspect ratios. This result is the same as what happens
in the case of flow between parallel plates. When one compares the amplitudes of the induced velocity, one finds that the amplitude of flow between flat plates is larger than
that in a square duct. This is due to the fact that, in a duct the fluid experiences friction
of four sides, whereas in the case of flow between parallel plates, it experiences flow only from two sides. When the aspect ratio is increased to a/h = 10, the velocity in the duct
differs only with the velocity between parallel plates near the side walls. This is clearly
due to the effects of viscosity.

For moderately pulsating frequencies, $ah = 8$, the velocity distribution of the flow in a duct
of square cross- sectional area differs considerably from that obtained at low frequencies.
The shapes of the velocity profiles are different; results indicate that, at certain instants of
time during a complete cycle, the profiles reach maximum values near the wall of the pipe
rather than on its axis of symmetry. This is Richardson's "annular effect". The induced velocity is no longer in phase, that is in lock step, with the driving pressure gradient. Rather,
the velocity is shifted with respect to the driving pressure and the magnitude of the shift depends on how far away points in the flow space are from the wall. Near the wall, the induced velocity on the axis of the duct lags behind that in the regions that are near the walls
of the duct. On the axis, the phase shift is 90°. This was true at low and at high aspect ratios.
This result is the same as what happens in the case of flow between parallel plates. When
one compares the amplitudes of the induced velocity, one finds that the amplitude of flow
between flat plates is larger than that in a square duct. This is due to the fact that, in a duct
the fluid experiences friction of four sides, whereas in the case of flow between parallel
plates, it experiences flow only from two sides. When the aspect ratio is increased to a/h =
10, the velocity in the duct differs only with the velocity between parallel plates near the
side walls. This is clearly due to the effects of viscosity.

5.2. Graphical illustration of the results of analysis by Yakhot, Arad and Ben-dor (1999)

The velocity profiles in pulsating flow at selected instants within one complete period are shown below. Flow in a duct is compared to flow between parallel plates for different aspect ratios and frequencies.

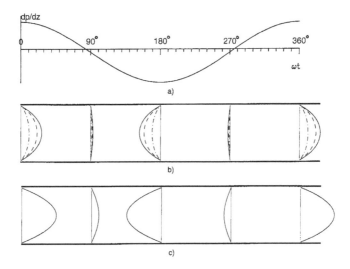

Figure 8. Velocity profiles in pulsating flow at different instants of one period. (a) Pressure gradient variation with time. (b) Duct flow, $a/h =1$, $a\ h =1$: solid line, $x/a = 0.5$; dashed, $x/a = 0.25$; dot-dashed, $x/a = 0.1$. (c) Flow between two parallel plates.

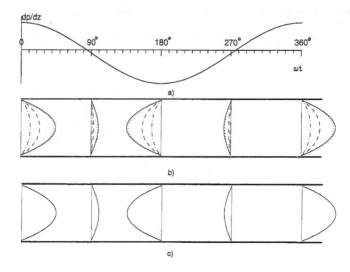

Figure 9. Velocity profiles in pulsating flow at different instants of one period. (a) Pressure gradient variation with time. (b) Duct flow, $a/h = 10$, $a\,h=1$: solid line, $x/a = 0.5$; dot, $x/a = 0.1$; dashed, $x/a = 0.025$; dot-dashed, $x/a = 0.01$. (c) Flow between two parallel plates.

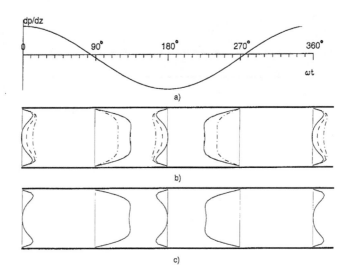

Figure 10. Velocity profiles in pulsating flow at different instants of one period. (a) Pressure gradient variation with time. (b) Duct flow, $a/h = 1$, $a\,h=8$: solid line, $x/a = 0.5$; dashed, $x/a = 0.25$; dot-dashed, $x/a = 0.1$. (c) Flow between two parallel plates.

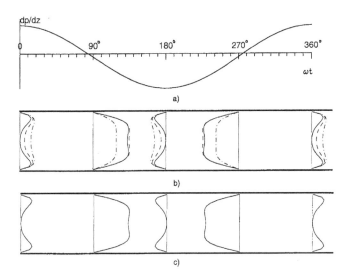

Figure 11. Velocity profiles in pulsating flow at different instants of one period. (a) Pressure gradient variation with time. (b) Duct flow, $a/h = 10$, $a h = 8$: solid line, $x/a = 0.5$; dashed, $x/a = 0.025$; dot-dashed, $x/a = 0.01$. (c) Flow between two parallel plates.

6. Anatomy of the shift using expansions of general results into power series

6.1. Series expansions of Kelvin functions

The unsteady part of the solution, which is given by $\frac{u'}{U}$, Eq. (22), can be written to show the pressure gradient explicitly as shown below.

$$\frac{u'}{U} = \sum_{n=1}^{\infty} W\left(r, a, k\right)\left\{\frac{\kappa_{cn}}{\kappa_0}\cos(nt - \varphi) + \frac{\kappa_{sn}}{\kappa_0}\sin(nt - \varphi)\right\} \tag{26}$$

Where

$$W(r, a, k) = \frac{8B}{(ka)^2}\left[B^2 + (1 - A)^2\right]^{1/2} \tag{27}$$

And $tan(\varphi(r, a, k)) = \frac{1 - A}{B}$

After a considerable amount of algebra using series expansions for the ber and bei functions, it can be shown that

$$W(r, a, k) = 2\left(1 - \frac{r^2}{a^2}\right) D(r, a, k) \tag{28}$$

Where D(r, a, k) is a dimensionless factor that is defined as shown below

$$D(r, a, k) = \left\{ \frac{\sum\limits_{n'=0}^{\infty} F_m(x, y)}{(ber^2 ka + bei^2 ka)} \right\}^{1/2} \tag{29}$$

Where $m = 4n'$, with $n' = 0,1,2,3,\ldots$, $x = ka$, $y = \frac{r}{a}$, and $F_m(x, y)$ denotes a family of polynomials a sample of which is shown below

$$F_0 = 1 \tag{30}$$

$$F_4 = \frac{4}{(4!)^2}\left(\frac{x}{2}\right)^4 (1 + 10y^2 + y^4)$$

$$F_8 = \frac{22}{(6!)^2}\left(\frac{x}{2}\right)^8\left(1 - \frac{14}{11}y^2 + \frac{186}{11}y^4 - \frac{14}{11}y^6 + y^8\right)$$

$$F_{12} = \frac{68}{(8!)^2}\left(\frac{x}{2}\right)^{12}\left(1 + \frac{66}{17}y^2 - \frac{277}{17}y^4 + \frac{948}{17}y^6 - \frac{277}{17}y^8 + \frac{66}{17}y^{10} + y^{12}\right)$$

$$F_{16} = \frac{254}{(10!)^2}\left(\frac{x}{2}\right)^{16}\left(1 + \frac{154}{127}y^2 + \frac{2206}{127}y^4 - \frac{10142}{127}y^6 + \frac{21610}{127}y^8 - \frac{10142}{127}y^{10} + \frac{2206}{127}y^{12} + \frac{154}{127}y^{14} + y^{16}\right)$$

$$F_{20} = \frac{922}{(12!)^2}\left(\frac{x}{2}\right)^{20}\left(1 + \frac{1066}{461}y^2 - \frac{2685}{461}y^4 + \frac{41964}{461}y^6 - \frac{158412}{461}y^8 + \frac{268476}{461}y^{10} - \frac{158412}{461}y^{12} + \frac{41964}{461}y^{14} - \frac{2685}{461}y^{16} + \frac{1066}{461}y^{18} + y^{20}\right)$$

$$F_{24} = \frac{3434}{(14!)^2}\left(\frac{x}{2}\right)^{24}\left(\begin{array}{l}1 + \frac{3238}{1717}y^2 + \frac{13040}{1717}y^4 - \frac{109654}{1717}y^6 + \frac{769653}{1717}y^8 - \frac{2359044}{1717}y^{10} + \frac{3530268}{1717}y^{12} - \\ - \frac{2359044}{1717}y^{14} + \frac{769653}{1717}y^{16} - \frac{109654}{1717}y^{18} + \frac{13040}{1717}y^{20} + \frac{3238}{1717}y^{22} + y^{24}\end{array}\right)$$

$$F_{28} = \frac{12868}{(16!)^2}\left(\frac{x}{2}\right)^{28}\left(\begin{array}{l}1 + \frac{25992}{12868}y^2 + \frac{24716}{12868}y^4 + \frac{337040}{12868}y^6 - \frac{2663036}{12868}y^8 + \frac{13416312}{12868}y^{10} - \frac{34632404}{12868}y^{12} + \frac{48192480}{12868}y^{14} - \\ - \frac{34632404}{12868}y^{16} + \frac{13416312}{12868}y^{18} - \frac{2663036}{12868}y^{20} + \frac{337040}{12868}y^{22} + \frac{24716}{12868}y^{24} + \frac{25992}{12868}y^{26} + y^{28}\end{array}\right)$$

Note, from the definition of w(r, a, k), Eq. (28), that each of these polynomials will be multiplied by the steady velocity. Clearly, this shows that all components that are added to the velocity due to unsteadiness are essentially various forms of the same steady velocity after it has been modified by the introduction of time variations. The series of equations shown below demonstrates this observation:

$$\frac{u}{U} = \frac{u_s}{U} + \frac{u'}{U}, \tag{31}$$

$$\frac{u_s}{U} = 2\left(1 - \frac{r^2}{a^2}\right), \tag{32}$$

$$\frac{u}{U} = 2\left(1 - \frac{r^2}{a^2}\right) + \sum_{n=1}^{\infty} W\left(r, a, k\right)\left\{\frac{\varkappa_{cn}}{\varkappa_0}\cos(nt - \varphi) + \frac{\varkappa_{sn}}{\varkappa_0}\sin(nt - \varphi)\right\},\qquad(33)$$

Using the expression for $W(r, a, k)$ that is shown in Eq. (28), one gets

$$\frac{u}{U} = 2\left(1 - \frac{r^2}{a^2}\right) + 2\left(1 - \frac{r^2}{a^2}\right)\sum_{n=1}^{\infty} D\left(r, a, k\right)\left\{\frac{\varkappa_{cn}}{\varkappa_0}\cos(nt - \varphi) + \frac{\varkappa_{sn}}{\varkappa_0}\sin(nt - \varphi)\right\},\qquad(34)$$

After a minor rearrangement of terms, Eq. (34) becomes

$$\frac{u}{U} = 2\left(1 - \frac{r^2}{a^2}\right) + 2\left(1 - \frac{r^2}{a^2}\right)\sum_{n=1}^{\infty} \frac{D(r, a, k)}{\varkappa_0}\left\{\varkappa_{cn}\cos(nt - \varphi) + \varkappa_{sn}\sin(nt - \varphi)\right\}.\qquad(35)$$

Since $D(r, a, k)$, in Eq. (35), consists of the functions $F_m(x, y)$, one concludes that the family of polynomials $F_m(x, y)$ that is shown in Eq.(30) is what is primarily responsible for the change in the shape of the velocity profile as the frequency of oscillation increases. Therefore, it is those polynomials that cause the location of the maximum velocity to move away from the axis of the pipe, and hence, bear the essence of the physical interaction between viscous forces and pressure forces during pulsating motions. This conclusion will be illustrated graphically below.

6.2. Graphical illustrations of the shape of the $F_m(x, y)$ polynomials

Variation in the shapes of the functions $F_m(x, y)$ is illustrated graphically below. It will be remembered that, in what follows, $x = ka = \sqrt{\frac{n}{v}}a$, $y = \frac{r}{a}$, the dimensionless radial distance from the axis of the pipe; and that

$m = 4n'$, with $n' = 1, 2, 3, 4, 5, \ldots$ In Figures 12 and 13, twelve functions $F_m(x, y)$ are plotted against the radial distance y for various values of the dimensionless parameter x.

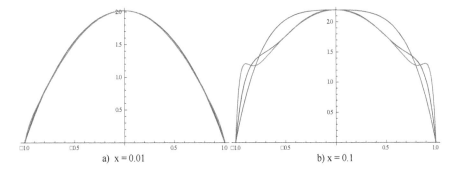

a) x = 0.01   b) x = 0.1

Figure 12. Each coordinate frame shows plots of three functions $F_m(x, y)$ vs y: $F_4(x, y)$, $F_{12}(x, y)$, and $F_{28}(x, y)$; x is used as the parameter. Note that larger values of x indicate higher rates of pulsations by the pressure gradient.

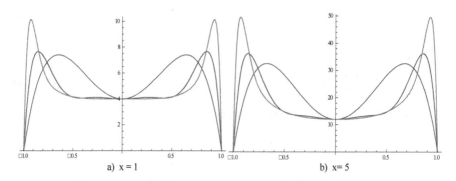

Figure 13. Each coordinate frame shows plots of three functions $F_m(x, y)$ vs. y: $F_4(x, y)$, $F_{12}(x, y)$, and $F_{28}(x, y)$; x is the parameter. Note that larger values of x indicate higher rates of pulsations by the pressure gradient.

7. Compiled summary of results from several investigators and conclusions

While conducting experiment on sound waves in resonators, Richardson (1928) measured velocities across an orifice of circular cross-section and found that the maximum velocity could occur away from the axis of symmetry and toward the wall. Sexl (1930) proved analytically that what Richardson observed could happen. Richardson and Tyler (1929-1930) confirmed these findings with more experiments with a pure periodic flow generated by the reciprocating motion of a piston. Uchida (1956) studied the case of periodic motions that were superposed upon a steady Poiseuille flow. An exact solution for the pulsating laminar flow that is superposed on the steady motion in a circular pipe was presented by Uchida (1956) under the assumption that that flow was parallel to the axis of the pipe.

The total mean mass of flow in pulsating motion was found to be identical to that given by Hagen-Poiseuille's law when the steady pressure gradient used in the Hagen-Poiseuille's law was equal to the mean pressure gradient to which the pulsating flow was subjected.

The phase lag of the velocity variation from that of the pressure gradient increases from zero in the steady flow to 90° in the pulsation of infinite frequency.

Integration of the work needed for changing the kinetic energy of fluid over a complete cycle yields zero, however, a similar integration of the dissipation of energy by internal friction remains finite and an excess amount caused by the components of periodic motion is added to what is generated by the steady flow alone.

It follows that a given rate of mass flow can be attained in pulsating motion by giving the same amount of average gradient of pressure as in steady flow. However, in order to maintain this motion in pulsating flow, extra work is necessary over and above what is required when the flow is steady.

Recently, Camacho, Martinez, and Rendon (2012) showed that the location of the characteristic overshoot of the Richardson's annular effect changes with the kinematic Reynolds number in the range of frequencies within the laminar regime. They identified the existence of transverse damped waves that are similar to those observed in Stokes' second problem.

All these results were obtained in flows through pipes of circular cross-sections and rectangular ducts. It is reasonable to expect that they would hold in the flow of air in a wind tunnel. Experimental results indicate that the Richardson's annular effect does occur in the test section of a subsonic wind tunnel. That behavior first appears unusual and, indeed, odd. However, as shown in this chapter, there is considerable experimental and analytical evidence in the literature that indicates that this behavior is due to high-frequency pulsations of the pressure gradient. Accordingly, in the case of a subsonic wind tunnel, it is probably due to the fast rate of rotation of fan blades. Indeed, in our wind tunnel, results from analysis and those from experiments differed only by about 5.7%.

Nomenclature of the symbols (with units)

α: a dimensionless ratio that combines the rate of pressure pulsations and the distance from the wall of the pipe;

η: a dimensionless distance from the wall of the pipe;

ρ: the mass density of the fluid (kg/m^3);

μ: the coefficient of absolute viscosity ($N.s/m^2$);

v: the coefficient of kinematic viscosity (m^2/s);

ω : denotes the circular frequency of pressure oscillations;

$\varkappa_{cn}$ and $\varkappa_{sn}$: Fourier coefficients of the pressure gradient; $\varkappa_0$ is the steady part of the pressure gradient (m/s^2);

a: the inside radius of a pipe through which an oscillating flow is moving;

c : the magnitude of a reference speed

G: the total mean value of the mass flow, U the total mean value of the velocity

i : the pure imaginary number; it is defined by $i^2 = -1$

J_0 : Bessel function of the first kind and of zero order

k: a dimensionless ratio used by Schlichting to denote the magnitude of the frequency of oscillation

K: a symbol used by Schlichting to indicate the magnitude of the pressure gradient

n: denotes the circular frequency of pressure oscillations (rad/s)

P: the pressure that drives the flow (N/m²)

$\frac{\partial p}{\partial x}$: the pressure gradient in the axial direction of an infinitely long pipe

r: the radial distance measured from the axis of the pipe (m)

R: the inside radius of a pipe of circular cross section (m)

t: time elapsed (s)

u: the axial velocity of the flow (m/s)

u_s: the steady part of the velocity u (m/s)

u' : the unsteady part of the velocity u (m/s

U : the mean speed (m/s) of the velocity u (m/s)

x: a dimensionless ratio that measures the rate of pulsations of the pressure gradient

y: a dimensional distance from the wall of the pipe (m)

Author details

Josué Njock Libii

Indiana University-Purdue University Fort Wayne, Fort Wayne, Indiana, USA

References

[1] Richardson, E.G. and Tyler, E. (1929). The transverse velocity gradients near the mouths of pipes in which an alternating or continuous flow of air is established. The Proceedings of the Physical Society, Vol. 42, part I, No. 231, pp. 1-15. ISSN 0370-1328.

[2] Ury, Josef F. (1964), A graphical method for a closer study of Richardson's annular effect, Zeitschrift fur angewandteMathematik und Physik (ZAMP) 15, number 3, pp. 306-311. ISBN/ISSN: 1420-9039 OCLC:43807374.

[3] Camacho, F.J.; Martinez, R.; Rendon, L. (2012) The Richardson's Annular effect and a transient solution of oscillating pressure-driven flow in circular pipes, eprint arXiv: 1207.1495.

[4] Sexl, T. (1930). Uber die von E. G. Richardson entdecktenAnnuraleffekt.Zeitschrift fur Physik, 61, 349-62. ISBN 0691114390.

[5] Uchida, S. (1956). The Pulsating viscous flow superposed on the steady laminar motion of incompressible fluid in a circular pipe. Zeitschrift fur angewandteMathematik und Physik 7, 403-422. ISBN/ISSN: 1420-9039 OCLC:43807374.

[6] Yakhot, A. Arad, M., and Ben-dor, G.(1999), Numerical investigation of a laminar pulsating flow in a rectangular duct, International Journal for Numerical Methods in Fluids, Vol. 29, Issue 8, pp 899-996, 30 April 1999. ISSN 0271-2091.

[7] NjockLibii, J. (2010) Laboratory exercises to study viscous boundary layers in the testsection of an open-circuit wind tunnel, *World Transactions on Engineering and Technology Education(WTE&TE)*, Vol. 8, No. 1, (March 2010), pp. (91-97), ISSN 1446-2257.

[8] JosuéNjockLibii (2011), Wind Tunnels in Engineering Education, in Wind Tunnels and Experimental Fluid Dynamics Research, Jorge Colman Lerner, UlfilasBoldes, editors, Chapter 11, InTech Publishers, 2011. ISBN 978-953-307-623-2.

[9] Yakhot, A., M. Arad, M., & Ben-Dor, G. (1998).Richardson's Annular Effect in Oscillating Laminar Duct Flows.*Journal of Fluids Engineering*, Vol. 120, 1, (March 1998) pp. (209-301), ISSN 0098-2202.

[10] Watson, G.N., (1944), A treatise on the theory of Bessel functions, Cambridge University Press, Cambridge, England, ISBN-10: 0521483913 | ISBN-13: 978-0521483919.

Diverse Engineering Applications

Chapter 6

Investigation of Drying Mechanism of Solids Using Wind Tunnel

Abdulaziz Almubarak

Additional information is available at the end of the chapter

1. Introduction

Drying of solids provides a technical challenge due to the presence of complex interactions between the simultaneous processes of heat and mass transfer, both on the surface and within the structure of the materials being dried. Internal moisture flow can occur by a complex mechanism depending on the structure of the solid body, moisture content, temperature and pressure in capillaries and pores. External conditions such as temperature, humidity, pressure, the flow velocity of the drying medium and the area of exposed surface also have a great effect on the mechanisms of drying.

Theoretical and experimental studies [1-6] reported the forced convection heat and mass transfer across different shapes. Most of these studies have considered a heated solid surface with a uniform surface temperature. However, this situation is not the same as in the drying process, where heat and mass transfer occurs simultaneously and the interfacial temperature and moisture content vary during the drying.

Evaporation from a flat plate surface to a laminar boundary layer was theoretically analyzed [7]. It was calculated distributions of the interfacial temperature and local Nusselt and Sherwood numbers for a parallel flow where both Prandtl and Schmidt numbers are unity. A conclusion stated that the characteristics of heat and mass transfer are highly conjugated and significantly influenced by the temperature dependency of vapor-liquid equilibrium, the magnitude of the latent heat of the phase change, and the thermal conductance of the flat plate. The work needs to be extended for the case of unsteady state conditions and to be repeated for a drying bed.

The variation in the surface temperature for a flat bed of a capillary porous material was discussed in [8]. The authors presented graphs that show an increase in surface temperature from the leading edge of the flat plate during the periods of drying. They also showed graphs in which the surface temperature decreased. In general, a physical meaning for that contradiction was needed to be considered.

As in [9], Jomaa *et al.* presented a simulation of the high-temperature drying of a paste in a scaled-up wind tunnel. They attempted to study the influence of the local air flow and thermal radiation on the drying behaviour of the product. A rapid air velocity (8 m s⁻¹) was used in the experiment, and an empirical model was derived to predict temperature and solvent content along the conveyer belt. Comparison of the experimental results with those predicted ones by the model showed a satisfactory agreement.

Evaporation of a pure liquid droplet has been widely studied both theoretically and experimentally [10-14]. However, in spray dryers and droplet drying applications, the droplets always consisted of multi-component mixtures of liquids and sometimes dissolved solids, forming a complicated multiphase composition. This makes analyzing heat and mass transfer processes more difficult. This effect is attributed to the presence of various components that vaporize at different rates, giving rise to a gradient in concentration in the liquid and vapor phase. In addition, a solid crust forms at the outer surface of droplet which acts as a resistance to heat and mass transfer processes.

Various experimental techniques have been used [15-19] to study the mechanisms of drying for a single droplet containing dissolved solids. The droplet was suspended freely from the end of a stable nozzle fixed in a wind tunnel. Air flow was hitting the droplet from one side causing a significant disturbance to the shape of droplet. Therefore, there was some difficulty in recording the weight and temperature of the droplet. The transferred heat conduction to the droplet by the nozzle was another problem.

Cheong *et al.* [20] proposed a mathematical model to predict core temperature for drying a free suspended droplet against time. Reasonable agreement was obtained between the predicted and the experimental results at an air temperature of 20ºC. However, the predicted temperature was less accurate at higher temperatures (50ºC and 70ºC); the model was not applicable for cases at high air temperatures.

A mathematical model was modified in [21], taking into account the droplet shrinkage. The droplet was assumed first to undergo sensible heating with no mass change. The model showed that temperature distribution within the droplet cannot be ignored even for a small diameter droplet of 200 μm.

Wind tunnel definitely is considered one of the best tools to investigate and to study the role of boundary layer and the mechanisms of drying process. The most important variables in any drying process such as air flow, temperature and humidity are usually easy to be controlled inside the wind tunnel. In the current study, through an experimental work and mathematical approach, we attempt to understand the role of the boundary layer on the interface behavior and the drying mechanisms for various materials of a flat plate surface and a single droplet shape.

2. Boundary layer over a flat plate surface

A boundary layer developed over a flat-plate surface plays a great role in the mechanisms of convective drying. Very little work has been done on the conjugated problem of heat and mass transfer during a flow over a drying bed. In this paper, through an experimental and mathematical approach, we attempt to understand the role of the thermal boundary layer on the interface behavior and the drying mechanisms for porous mediums. Beds of desert sand, beach sand and glass beads were subjected to forced convective drying in a scaled-up wind tunnel.

A laboratory-scale dryer designed for this work is shown in Figure 1. The apparatus consists of wind tunnel, molecular-sieve air dryer, 3 KW air heater and a fan. Through the wind tunnel a controlled flow of hot, dry air, with an average velocity of 1 m s^{-1}, was passed over a sample mounted flush with the tunnel floor. The last section of the wind tunnel (converging section) was designed to be opened easily for installation of the test bed.

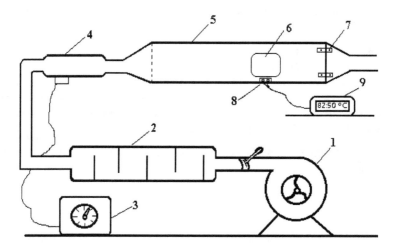

Figure 1. Experimental apparatus composed of: 1. fan; 2. molecular-sieve air dryer; 3. voltage regulator; 4. air heater; 5. wind tunnel; 6. observation port; 7. joint; 8. thermocouple socket; 9. thermocouple selector.

A glass tray (100 cm x 36 cm x 2.5 cm) attached to a flat metal plate was especially designed for this study. The top surface of the glass tray was at the same level of the metal plate that formed the floor of the wind tunnel. The sides and the bottom of the tray are insulated with neoprene rubber to minimize heat transfer *via* the glass wall. Figure 2 shows both the wind tunnel and the glass tray.

Ten thermocouples were inserted from one side of the tray to facilitate measurement of surface temperature distribution along the bed. At the same side of the tray, another ten thermocouples were inserted but at lower depth, 2.2 cm from the surface, to measure the bottom temperature distribution.

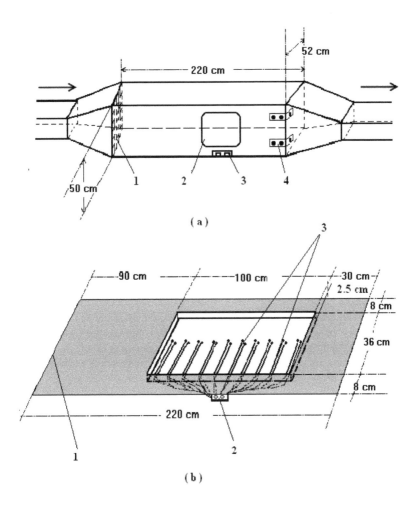

Figure 2. Details of (a) Wind tunnel: 1. smoothing grid; 2. observation port; 3.thermocouple socket; 4. joint, (b) Glass tray (clear) and flat plate (gray): 1. the leading edge; 2. thermocouple socket; 3. thermocouples.

The experiment was initiated by switching on the centrifugal fan and then the electric air heater. The voltage regulator was adjusted to provide the desired air temperature. The air temperature was monitored until it reached a steady state. This state was normally required 1 - 2 hours to be achieved.

When the apparatus achieved a constant air temperature, the drying process was commenced. The metal plate and the glass tray, containing the sample, were placed carefully inside the tunnel, allowing the hot air to pass over the surface of the bed. The initial readings of time and temperature were then registered immediately. During the experiment, temperature distributions were measured at intervals of approximately 20 minutes. The temperature at each distance was measured by using the thermocouple selector and registered with an accuracy ± 0.05ºC.

Two types of sand were subjected to the drying process. The first type was a fine sand of average grain size 220 μm diameter, with a moisture content of 0.17 kg kg^{-1}, taken from the desert of Kuwait, Burgan (N 28° 44 00 E 047° 42 00). The nature of the desert sand is usually fine and dry. The other type of sand was a beach sand of average grain size 300 μm diameter and 0.24 kg kg^{-1} moisture content, taken from the north coast of Kuwait, Bobiyan Island (N 29° 46 00 E 048° 22 00). The north coast is a place where the Shatt Al-Arab River (Iraq) falls into the Arabian Gulf. This makes the coast more muddy and less saline than the south coast. In addition, the beach sand contains a lot of small shells of various shapes. Glass beads of 400 μm diameter and 0.2 kg kg^{-1} initial moisture content were also selected, for test as a porous medium for comparison. The bed of glass beads is considered a typical sample for drying experiments, since the sizes of the beads are almost identical. The two types of sand and the glass beads beds are good examples for testing drying processes.

2.1. Results

A bed of the desert sand was subjected to forced convective drying at 84 °C. The wet-bulb temperature of the air was 35.5 °C. The resulting surface temperature vs. time at different distances from the leading is shown in Figure 3, which demonstrates clearly the stages of the drying process; i.e. the pre-constant rate period, the constant rate period, and the falling rate period.

The temperature of the bed rose from ambient temperature 23°C to a steady value at time = 70 minute. This initial period, termed the pre-constant rate period, is usually short. The surface temperatures remained constant for a period of 140 minutes, indicating the constant rate period. The surface temperature at the distance x=1cm, from the leading edge was 36 °C. This temperature was greater than that at x=50 cm and at x= 100 cm by 1 °C and 2 °C respectively. The surface temperatures were close to the wet-bulb temperature.

During the constant rate period, the surface of the solid is so wet that a continuous film of water exists on the drying surface. This water is entirely unbound water and exerts a vapour pressure equal to that of pure water at the same temperature. The rate of moisture movement within the solid is sufficient to maintain a saturated condition at the surface.

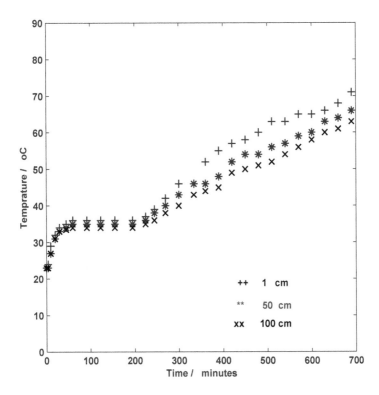

Figure 3. Temperature distribution profile of the surface for the desert sand bed.

At a specific point, t = 240 minutes, the surface temperature at all positions rose gradually, indicating the end of the constant-rate period and the beginning of the falling-rate period. In the falling-rate period, there was insufficient water on the surface to maintain a continuous film. The entire surface was no longer wetted and dry patches began to form. The surface temperature continued to rise for a longer time until it approximated to the air temperature. A thin dried layer appeared on the entire surface.

The temperatures at the surface and the bottom at different distances from the leading edge are shown in Figure 4. The profiles show that when the surface became dry, the bottom remained wet at a constant temperature for a longer time than that for the surface. During the falling-rate periods a receding evaporation front divided the system into a hotter, dry zone near the surface and a wet zone towards the bottom of the sample [22]. The evaporation plane receded from the surface toward the bottom. The temperatures then rose quickly when the dry zone extended throughout the bed.

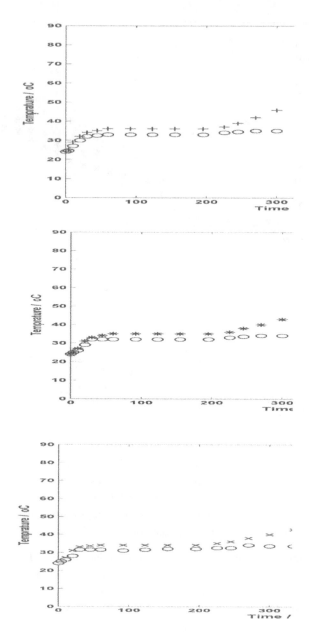

Figure 4. Temperatures of surface and bottom for the desert sand bed at distance of 1 cm, 50 cm and 100 cm from the leading edge, respectively, from the top.

A bed of the beach sand was dried at an air temperature of 83 °C. The wet bulb temperature was 35 °C. The temperature distribution profile at different distances from the leading edge is shown in Figure 5. The stages of the drying process can easily be recognized from the temperature profile. However, the surface temperature took longer to approach to the air temperature.

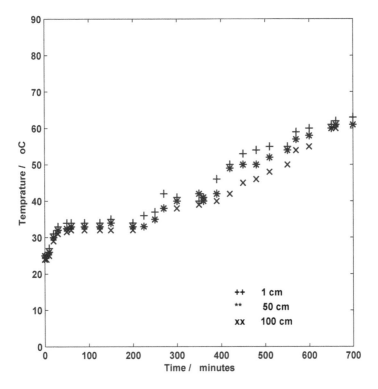

Figure 5. Temperature distribution profile of the surface for the beach sand bed).

Figure 6 shows the surface and the bottom temperatures at distances from the leading edge of 1 cm, 50 cm 100 cm, respectively. It can be seen that the surface and bottom temperatures increased rapidly at some times and decreased others. Also, the temperature profile did not increase gradually like that of the desert sand. This can be attributed to the nature of the beach sand, which comprises different types of small shells of various shapes, and contain tiny hollows. The trapped water in these hollows forms small bubbles which can explode with increasing temperature. Therefore the temperature of the bed changes suddenly at such times. The temperature at the bottom of the bed indicated it remained wet during the falling-rate period.

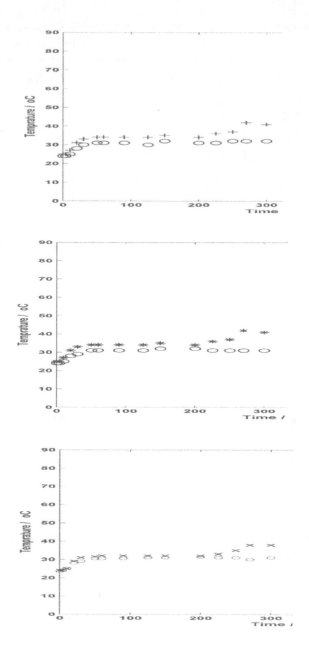

Figure 6. Temperatures of surface and bottom for the beach sand bed at distance of 1 cm, 50 cm and 100 cm from the leading edge, respectively, from the top.

A similar investigation was carried out for a bed of the glass beads at an air temperature of 84 °C. The wet bulb temperature was 37 °C. Figure 7 shows temperature versus time at different distances from the leading edge of the bed surface. The temperature distribution profile again illustrates clearly the stages of the drying process. The results for the surface and bottom temperatures for the glass beads were almost similar to those for the desert sand as shown in Figure 4.

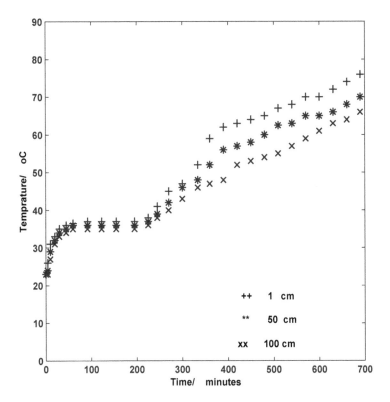

Figure 7. Temperature distribution profile of the surface for the glass-bead bed.

2.2. Discussion

A thin film adjacent to the surface always exists when a forced flow passes over a flat plate and forms what is called a hydrodynamic boundary layer. The influence of the surface temperature reaches deeper into the fluid, thus causing the formation of a thermal boundary layer. It is well known, the thickness of the thermal boundary layer increases with increasing distance from the leading edge. This layer is affected by the geometry of the system, roughness of the surface and the fluid properties.

For the case in which the heated section is preceded by an unheated straight length, the local Nusselt number (Nu_x) is represented in [23-25] as:

$$Nu_x = \frac{hx}{k} = \frac{0.323(Re_x)^{0.5}(Pr)^{1/3}}{[1-(x_o/x)^{3/4}]^{1/3}} \tag{1}$$

where Re_x is Reynolds number with respect to length and x is the length of the flat plate in (m).

Figure 8 shows a plot of variation in the local heat transfer coefficient versus the distance from the leading edge. The plot indicates that the values of the coefficient decreased significantly when x increased from the leading edge, and then it remains virtually constant for large x values. A plot of variation of the mass transfer coefficient could be expected to be almost similar to that in Figure 8 because of the similarity in the transport coefficient equations. This result with the concept of the boundary layer thickness demonstrates that resistance to heat and mass transfer to, or from, the bed increases with increasing the distance from the leading edge.

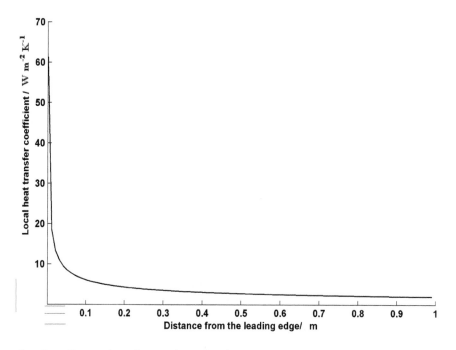

Figure 8. Local heat transfer coefficient vs. distance from the leading edge.

A model proposed in a previous paper [26] was modified to find a method for prediction of surface temperature distribution. The equation of energy can be represented as:

$$\varrho \ cp \ \frac{\partial T}{\partial t} = \nabla(k \nabla T) + \nabla(\frac{D_v M}{R T} \nabla P_v) \Delta hv \tag{2}$$

At time zero, the whole body has a uniform initial temperature of T_o, and the initial conditions are:

$$T_0 = T\big|_{y=0} = T\big|_{y=h} \tag{3}$$

At the external surface, i.e. y=0, the boundary conditions can be written on the basis of Figure 9 as:

$$-k \ \frac{\partial T}{\partial y} = h \ (T_a - T_{sx}) + \overset{\bullet}{m} \ \Delta hv \tag{4}$$

where T_{sx} is the surface temperature at distance x from the leading edge.

The drying rate, $\overset{\bullet}{m}$ varies with the time and can be defined as,

$$\overset{\bullet}{m} = \frac{M K'_G}{R T} (P_{sr} - P_v) \tag{5}$$

where K'_G is an overall mass transfer coefficient, defined in [27] as:

$$\frac{1}{K'_G} = \frac{1}{k'_c} + \frac{z''}{D_{eff}} \tag{6}$$

where k'_c is the local mass transfer coefficient (m s^{-1}), z'' is the distance from the plate surface to the receding evaporation front in (m).

The boundary condition at $y = h$ is:

$$-k \ \frac{\partial T}{\partial y} = 0 \tag{7}$$

Equation (2), with such boundary conditions, was solved by a finite difference method (a modified form the so-called explicit method). Therefore, the temperature distribution on the surface (i.e. $y = 0$) was calculated by using the model at different local points, x. Table 1 shows the physical properties of the sample materials.

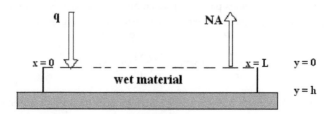

Figure 9. Heat and mass transfer process for the wet material. NA: mass transfer flux and q: heat transfer flux.)

Figure 10 shows the experimental and the predicted surface temperature distributions for the desert sand along the bed at various times. The predicted temperatures were in good agreement with the experimental results. From the graph, it can be seen that there is a significant difference in the surface temperature between 0.1 m and 1 m. At time=195 minutes; i.e. during the constant-rate period, the difference in temperature was 2 °C.

During the falling-rate period, the difference in the surface temperature between 0.1 m and 1 m increased. At 510 minutes, the difference is 11 °C. At the bottom of the sand bed, the difference in temperature along the bed also can be seen clearly (Figure 4).

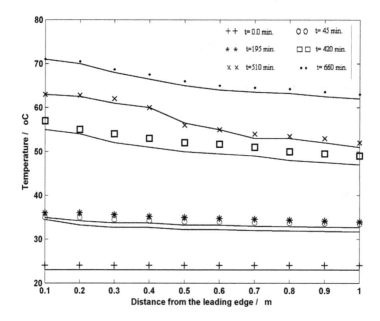

Figure 10. Experimental and predicted surface temperatures of the desert sand bed; experimental results (symbols); predicted results (solid lines).

For the entire time of the experiment, the surface and the bottom temperatures decreased gradually with increasing distance from the leading edge. This caused by the resistance to heat transfer process which increased with increasing thickness of thermal boundary layer. In contrast, near the leading edge, the resistance to heat transfer diminishes, since the thickness of the thermal boundary layer in the vicinity of the surface thins. Therefore, the rate of heat transfer to the body increased, thereby raising the temperature of surface. Afterwards, heat transfer by conduction across the solid particles raises the temperature of the bed, and the portion closest to the leading edge dries faster than that at a greater distance.

Figure 11 shows both predicted and experimental results for the bed of beach sand. This Figure shows that the computed temperature distribution was in general agreement with experimental results. However, unsatisfactory results can be seen at times greater than 400 minutes, i.e. during the falling-rate period. This is due to the nature of the beach sand as discussed before. The surface temperature of the beach sand sample decreased gradually with increasing distance from the leading edge. A significant difference in temperature between 0.1 m and 1 m can be seen clearly for both constant and falling-rate periods. During the constant-rate period, the difference in temperature was 2 °C, whereas in the falling-rate period the difference reached to 8 °C.

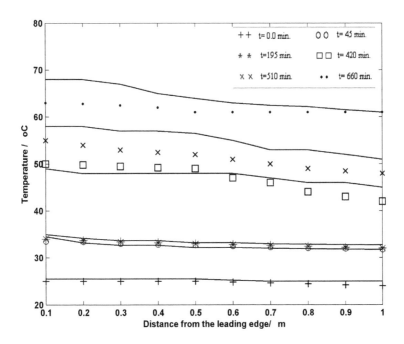

Figure 11. Experimental and predicted surface temperatures of the beach sand bed; experimental results (symbols); predicted results (solid lines).

	ρ(Kg/m³)	cp$_d$ (kJ/kg K)	k (W/m K)	k$_d$ (W/m K)	D$_{eff}$ (m²/s)	D$_v$ (m²/s)
Desert sand	1760	1.1	0.760	0.246	0.81x10^{-5}	1.75x10^{-5}
Beach sand	1980	1.2	1.12	0.188	1.15x10^{-5}	2.05x10^{-5}

Table 1. Physical properties for desert sand and beach sand.

We have found that the temperature distribution profiles determined for the flat beds of desert sand, beach sand and glass beads identified clearly the stages of drying. The temperature profiles in general were almost similar. However, the beach sand profile showed irregularity in temperature, due to the nature of the beach sand which contains a lot of shells with various shapes.

The temperature profiles also showed that when the whole surface of the bed became dry, the whole bottom of the bed remained wet. During the falling-rate periods, a receding evaporation front divided the system into a hotter, dry zone near the surface and a wet zone towards the bottom of the sample.

The predicted transport coefficients have very large values close to the leading edge, where the thickness of the boundary layer approaches zero. In contrast, the values of the coefficients decrease progressively with increased distance from the leading edge, where the boundary layer thickens. Hence the resistance to heat and mass transfer to, or from, the surface also increases. These variations in thickness and resistance have a significant effect on the temperature distribution along the bed and the drying rate.

A mathematical model has been modified to predict temperature distributions along the bed at various times. The model was compared with the experimental results for various beds and good agreement was obtained. We found that surface and bottom temperatures decreased gradually with increasing distance from the leading edge, and the difference in temperature became clearer during the falling-rate period. The difference in the surface temperature was 11 °C for the case of desert sand, and was 8 °C for the case of beach sand. We concluded that the portion close to the leading edge dried faster than that at larger distance, since the resistance to heat and mass transfer diminishes at that position.

3. Single droplet drying

The study of mechanisms that describe single droplet drying is a challenging issue since it involves many disciplines: heat and mass transfer, fluid mechanics and chemical kinetics. The main objective of this section is to attempt to further understand the mechanisms involved in the drying of a single droplet, and more specifically to formulate a mathematical model. The model should predict temperature profiles for both the inner core and the outer surface of the droplet under simulated conditions that might be encountered in spray drying equipment. Experimental work will also involve the measuring the moisture contents against time to provide more information for the droplet drying process.

The experimental apparatus was comprised of a horizontal wind tunnel 2.2 m long. The wind tunnel supplied a forced drying air into the working section where the droplet was suspended from a glass nozzle. The apparatus and the flow system are shown in Figure 12. A gate valve at the inlet to the wind tunnel controlled airflow rate. Air was heated to the desired temperature, using a 3 KW electric heating element controlled by a rotary voltage regulator.

Through the wind tunnel, a controlled flow of hot, dry air, with an average velocity of 1 m/s, was passed across the droplet suspended from the glass nozzle. A forced dry air was obtained by using a centrifugal fan and a molecular sieve air dryer containing silica gel and calcium silicate.

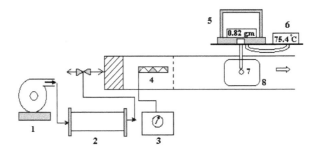

Figure 12. Experimental apparatus: 1-fan 2-molecular sieve 3-voltage regulator 4-air heater 5-Digital balance 6-Temperature recorder 7-Glass nozzle 8-Observation port.

The experiment was initiated by switching on the centrifugal fan and then the electric air heater. The voltage regulator was adjusted to provide the desired air temperature. A thermocouple was fixed in the wind tunnel near to the glass nozzle to measure the air temperature with an accuracy of $\pm 1°$ C. The air temperature was monitored until it reached a steady state. This state requires between 1 to 2 hours to be achieved. When the apparatus achieved a constant air temperature, the drying process was initiated.

As a part of this study, a droplet suspension device was specially designed to measure the weight and temperature of the droplet. The droplet suspension device is illustrated in Figure 13. It consisted of a glass nozzle with the dimensions of 180 mm in length and 9 mm outside diameter. The upper section of the glass nozzle was fixed by a small electric rotator device able to provide a rotational speed range of 1- 30 rpm. In order to reduce the contact area between the free end of the glass nozzle and the droplet surface, the lower section of the nozzle was shaped as a small cone with a free end diameter of 4 mm. The droplet receives a relatively small amount of heat transferred by conduction through the glass nozzle, which has a thermal conductivity of 0.480 W/m K. The heat transferred by conduction was taken into account in the proposed model represented by Equations 4&5. The glass nozzle was rotated at constant low speed of 5 rpm. This rotation speed had no effect on the shape or stability of the droplet. The upper section of the suspension device was fixed by a metallic clip installed beneath the analytical balance. The lower section of the suspension device, i.e. the glass nozzle, was inserted through a hole in the wind tunnel.

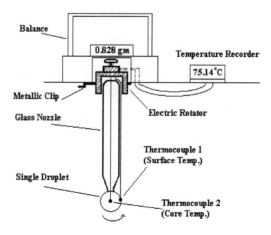

Figure 13. Droplet suspension device; glass nozzle and the connected thermocouples.

Two thermocouple sensor types, NiCr-NiAl, were used to measure the core and surface temperatures of the droplet. One of the thermocouples was placed inside the glass nozzle and extended to the center of the droplet. The other thermocouple was fixed outside and along the glass nozzle. The outer thermocouple rotates simultaneously with the rotation of the glass nozzle. The end tip of the thermocouple was positioned in a manner to touch the outer surface of the droplet. The rotation process that made both the nozzle and the droplet rotate together assisted in avoiding any separation between them that might have been caused by the force of air drying. The core and the surface temperatures of the droplet were easily recorded by a temperature recorder at 50 sec. intervals. The drying process of the different material droplets was investigated under air temperature of 75° C and 140° C.

The procedure for weighing the droplet was carried out quickly and intermittently by causing the suspension device to be freely-suspended. The gate valve was closed to cut off the airflow to the working section and diverted to an outlet 20 mm valve in order to prevent any vibration of the glass nozzle during the weighing process. The nozzle rotation was simultaneously stopped. A metallic clip was opened manually to allow the suspension device to be freely-suspended from a hook connected beneath the balance. This arrangement made the weight measurement readings more accurate. The weighing procedure was repeated during the droplet drying experiment at 100 secintervals. Thus, the weight loss of droplet was recorded and moisture content was determined versus time. The required time for each weighing procedure step was about 10 sec. This time was not included in the recorded drying time and had no noticeable effect on the results.

Three types of liquids were selected for the drying process experiments. The first type was sodium sulphate decahydrate solution (60 wt % solid). The second type was a concentrated fruit juice (60 wt % fruit juice powder of apple, peach and blueberry, MTC product). The third type was an organic paste (20% sodium chloride, 25% dispersal pigment; Goteks product) used for adhesive and coating applications. Table 2 shows the physical properties of the sample materials. Droplets ranging from 9 mm to 14 mm diameter were subjected to the drying process. However, the actual size of the droplets in typical spray drying applications is much smaller. The droplets with small sizes require developing a more accurate technique to measure both core and surface temperatures. Therefore, the current research assumes that the mechanisms of the drying large and small droplets are similar.

	ρ_d (Kg/m³)	cp_d (kJ/kgK)	k (W/mK)	k_d (W/mK)	D_{eff} (m²/s)	D_v(m²/s)
Sodium sulphate decahydrate solution	3110	1.1	0.180	0.246	1.14x10⁻⁵	3.45x10⁻⁵
Fruit juice	1650	1.7	0.126	0.188	1.10x10⁻⁵	3.40x10⁻⁵
Organic paste	3400	2.4	0.251	0.422	1.52x10⁻⁵	3.50x10⁻⁵

Table 2. Physical properties of sodium sulphate decahydrate solution, fruit juice and organic paste.

3.1. Mathematical model

In the drying process, the droplet is first heated by the hot air flow with significant evaporation from the surface. The temperature of the surface increases and approaches the wet-bulb temperature, indicating the constant drying rate period. In order to propose a mathematical model, the droplet was assumed to have a fixed size with no change during this period. Also, the droplet was assumed to have a uniform initial temperature and moisture content. Temperature distribution within the droplet can be represented as

$$\frac{\partial T}{\partial t} = \alpha \left(\frac{\partial^2 T}{\partial r^2} + \frac{2}{r} \frac{\partial T}{\partial r} \right) \tag{8}$$

Equation 8 was solved with the following boundary conditions using explicit finite differences,

$$-k \frac{\partial T}{\partial r} = 0 \quad at\ r = 0 \tag{9}$$

and

$$-k \frac{\partial T}{\partial r} = h(T_a - T_{sr}) + q_{nz} \quad at\ r = R \tag{10}$$

where, q_{nz} is the transferred heat conduction to the droplet through the glass nozzle. It can be calculated as:

$$q_{nz} = (\frac{4 k_{nz} h_{nz}}{d_{nz}})^{0.5} (T_a - T_{sr})$$

(11)

The heat transfer coefficient, h_{nz}, can be correlated from Thomas (1999) as:

$$h_{nz} = 0.26 + Re^{0.6} Pr^{0.33} (k_{air} / d_{nz})$$

(12)

During the falling rate period, the formation of a receding evaporation front divides the droplet into two regions, a dry crust at the outer surface and a wet region inside the core. Therefore, heat transfer equations are formulated for each region. The physical model and the coordinate system for analysis are shown in Figure 14.

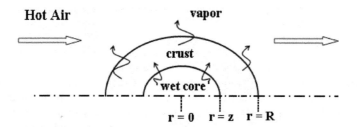

Figure 14. Physical model and the coordinate system of the droplet cross-section.

Energy balance for the wet core, $0 < r < z$, can be represented as follows:

$$\frac{\partial T_w}{\partial t} = \alpha_w (\frac{\partial^2 T_w}{\partial r^2} + \frac{2}{r} \frac{\partial T_w}{\partial r})$$

(13)

Heat is transferred through the crust into the wet core where evaporation occurs at the interface between the core and the crust. Vapor then diffuses through pores of the crust into the drying medium. Thus, moisture is transferred mainly by vapor flow. Consequently, vapor diffusion must be taken into account in formulating the equations for the dry (crust) region. The energy balance for the crust region, $z < r < R$, can be represented as follows:

$$\frac{\partial T_d}{\partial t} = \alpha_d (\frac{\partial^2 T_d}{\partial r^2} + \frac{2}{r} \frac{\partial T_d}{\partial r}) + \frac{\partial}{\partial r} (\frac{M}{RT} \frac{\partial P_v}{\partial r}) \frac{\Delta h_v}{\rho_d c p_d} D_v$$

(14)

3.2. Boundary conditions

At the center of the sphere, r = 0

$$-k_w \frac{\partial T_w}{\partial r} = 0 \tag{15}$$

At the surface, r = R:

$$-k_d \frac{\partial T_d}{\partial r} = h(T_a - T_{sr}) + q_{nz} \tag{16}$$

At the receding evaporation front (r = z), the moving boundary conditions are

$$-k_d \frac{\partial T_d}{\partial r} + k_w \frac{\partial T_w}{\partial r} = \overset{\bullet}{m} \, \Delta hv \tag{17}$$

The drying rate, $\overset{\bullet}{m}$, was defined in Eqs. (5 & 6). However, in this case, z'', in Eq. (6) represents the distance from the droplet surface (r = R) to the receding evaporation front (r = z).

Heat and mass transfer coefficients, h and k_c', can be determined by the correlations found in [28,29] as follows:

$$Nu = 2 + \Phi Re^{0.5} Pr^{0.33} \tag{18}$$

and

$$Sh = 2 + \beta Re^{0.5} Sc^{0.33} \tag{19}$$

where Φ and β are constants ranging from 0.6-0.7 for Re (500 -17000).

Thermal conductivity of the wet region can be evaluated according to [30] as:

$$k_w = k_d + k_v \overline{X} \tag{20}$$

where $\overline{X}$ is an average moisture content and k_v is

$$k_v = \frac{D_v M}{RT} \cdot \frac{dp}{dT} \Delta h_v \tag{21}$$

The non-linear Eqs. (13&14) with the boundary conditions as in Eqs. (15-17) were solved by a finite difference method. The proposed equations were solved in a program using Turbo-Pascal V.6, and the computational results compared with the experimental results.

3.3. Results

A single droplet of sodium sulphate decahydrate solution was suspended from the free end of the glass nozzle at an air temperature of 75° C and an air velocity of 1 m/s. The wet-bulb temperature was 46° C. The core and surface temperatures of the droplet versus time are plotted in Figure 15. Initially, the temperatures increased rapidly because of the large difference in temperature between the drying medium and the droplet. Basically, the plot exhibited two drying periods. A short period, from time =100 - 200 s, where the surface temperature approached the wet-bulb temperature, is described as the constant rate period. The second period, or falling rate period, extended from t =200 s to the end of the experiment.

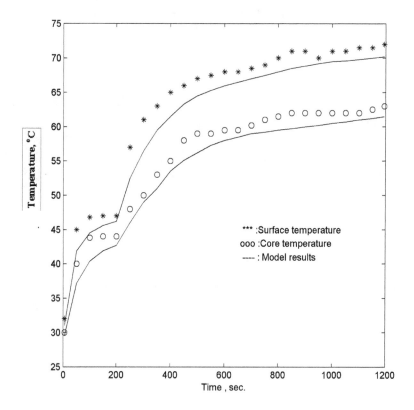

Figure 15. Temperature distribution profile for sodium sulphate decahydrate at air temperature of 75° C.

Figure 15 plots the predicted values of both the core and surface temperature evaluated by the developed model. Comparison of the experimental and theoretical results showed good agreement. However, actual surface temperatures were slightly higher than those predicted by the model. This probably was due to the position of the thermocouple and its reading during the experiment, as will be discussed later in more detail.

The previous experiment was repeated at an air temperature of 140° C in the same drying medium. Figure 16 shows experimental and theoretical results of a droplet of the same material. The experimental results showed that the constant rate period extended from time= 85 - 165 s, shorter than that observed at 75° C. The results showed both surface and core temperatures of 130° C at 900 s, at which point the droplet had dried completely. Figure 16 showed agreement between the experimental and predicted results.

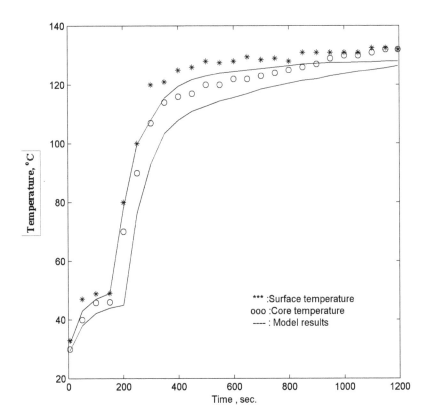

Figure 16. Temperature distribution profile for sodium sulphate decahydrate at air temperature of 140° C.

A single droplet of a fruit juice was dried at an air temperature of 75° C and an air velocity of 1 m/s. Wet-bulb temperature was 44.5° C. The experimental results in Figure 17 show a constant rate period from time =100 - 200 s. The falling rate period can be observed after the constant period, when the temperature increased quickly.

During the falling rate period some experimental readings represented approximate values of the actual droplet surface temperatures. The approximate values can be attributed to the fact that the droplet diameter constantly decreased with time and that caused the distance between the tip of thermocouple and the surface of droplet to grow. In other words, during some time in the experiment, a part of the end tip of the thermocouple was touching the surface of the droplet, and the remaining area of the tip was exposed to air flow. Such a behavior caused the tip of thermocouple to give an average reading for both the surface and the air flow temperatures. Good agreement was obtained between the theoretical and experimental results of the temperature profile for the fruit juice.

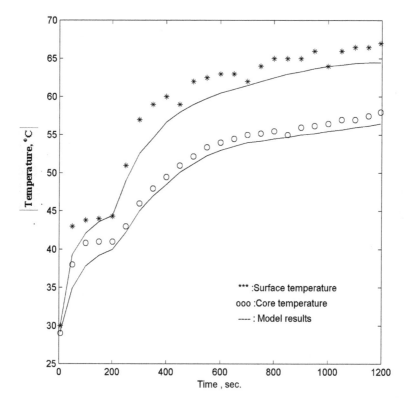

Figure 17. Temperature distribution profile for fruit juice at air temperature of 75° C.

The fruit juice droplet was dried again at an air temperature of 140° C in the same drying medium. The results (Figure 18) showed that the constant rate period was a little shorter than that at 75° C. The results showed also that both surface and core temperatures were similar and close to the air temperature at time = 1050 s. The mathematical model showed also good agreement with the experimental readings.

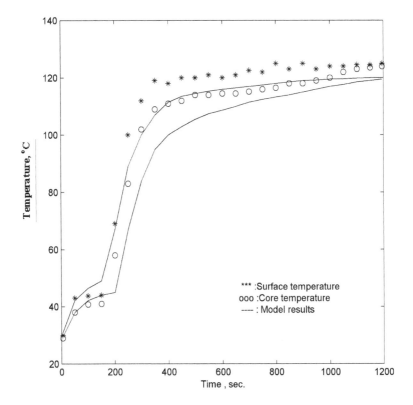

Figure 18. Temperature distribution profile for fruit juice at air temperature of 140° C.

A single droplet of organic paste was dried under similar conditions to that of the fruit juice. A plot of the core and surface temperatures of the droplet versus time is shown in Figure 19. A relatively longer period of constant drying rate, compared to those shown for sodium sulphate and fruit juice, was observed. Figure 19 also showed that surface temperature increased rapidly after time = 300 s, indicating the beginning of the falling rate period. Less agreement was obtained between the model's predicted results and the experimental results. Organic paste, which has higher thermal conductivity, forms a thicker solid crust at the outer surface compared to the other material evaluated. Therefore, a higher resistance to heat and vapor through the crust would be expected. To improve the predictions, introduction of a correction factor is probably required in the model for those materials which have a nature similar to that of the organic paste. Figure 20 shows the experimental and theoretical results of air temperature of 140° C. Again, forming a solid crust at the outer surface led to less agreement between theoretical and experimental results. The predicted temperature distributions were higher than the actual temperatures.

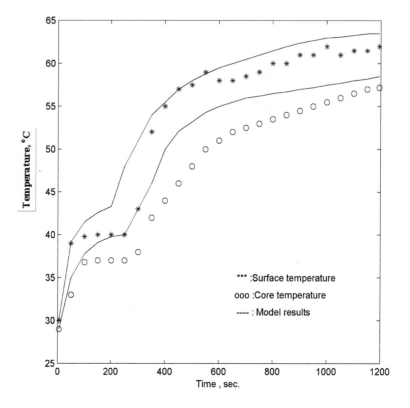

Figure 19. Temperature distribution profile for organic paste at air temperature of 75° C.

The moisture content of the droplet was determined by measuring the weight loss against time. There was no experimental technique to measure core and surface moisture content separately; therefore, the measured moisture content of the droplet represented the average value. The experimental results of moisture content distribution for the three samples at an air temperature of 75° C, are shown in Figure 21. The moisture content profiles clearly show the two stages of drying, the constant rate period and the falling rate period. The profiles show that the sodium sulphate decarbohydrate solution had consistently lower moisture content, as it dried faster than the other materials. The profiles also showed that forming a solid crust in the falling rate period lowered the moisture content values for all the three samples. In the case of organic paste, the change in the moisture content was more significant.

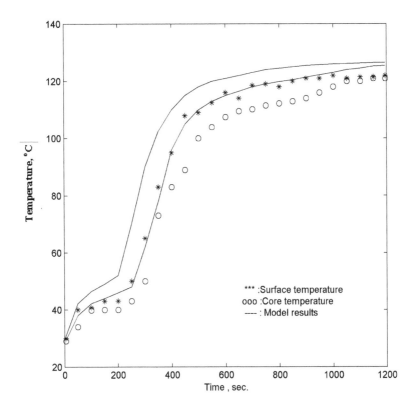

Figure 20. Temperature distribution profile for organic paste at air temperature of 140° C.

Obviously, crust formation, thickness and porosity have a significant effect on the moisture content and on the drying rate of the droplet. This result was also obtained by Hayder & Mumford [31] in the drying of custard and starch droplets. They observed that crust formation was more rapid on the custard droplet, because the smaller starch granules absorbed less and left more free water in the droplet. Therefore, the crust growth and the drying rate of the starch droplets took a longer time.

Moisture distribution curves for the three samples at 140° C air temperature are plotted in Figure 22. Similar results were obtained to those observed at an air temperature of 75° C. Also, the profiles showed that the moisture content dropped to lower values compared to those at 75° C. In other words, the droplets dried faster at the higher temperature.

3.4. Discussion

As previously observed, moisture and temperature distribution profiles of various materials exhibited constant and falling drying rate periods. In the constant rate period, the tempera-

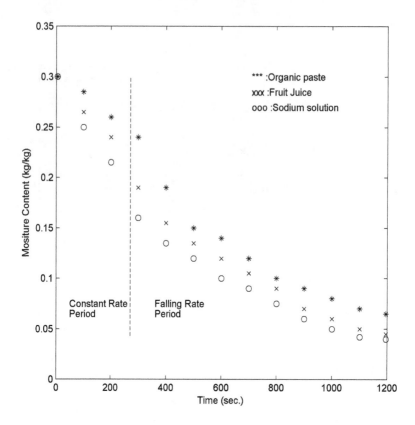

Figure 21. Moisture content profile for all samples at air temperature of 75° C.

ture of the droplet surface was almost equal to the wet-bulb temperature. During this period, evaporation takes place from the free liquid surface of the droplet. The constant rate period was relatively short in sodium sulphate and fruit juice samples. However, that period was longer in the case of organic paste.

The falling rate period is characterized by formation of a partial crust on the outer surface of the droplet. This crust recedes towards the core and the surface temperature starts to increase. Vapor diffusion becomes the predominant transport process at this stage. Crust structure, thickness and porosity have a significant effect on the rate of drying. The crust thickness increases with time, hence the resistance to heat and moisture diffusion through the crust increases. Therefore, moisture content and drying rate decrease.

Some experimental readings represented approximate values of the actual droplet surface temperatures. This was attributed to the end tip of thermocouple that was giving average readings for both the surface and the air flow temperatures at the same time. It was also as-

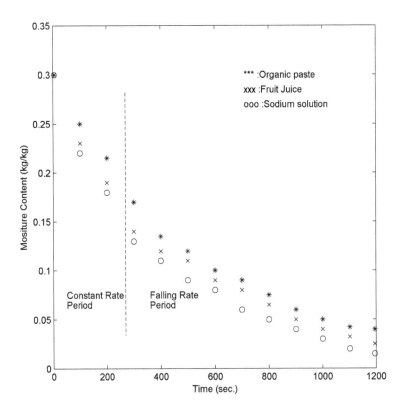

Figure 22. Moisture content profile for all samples at air temperature of 140° C.

sumed that the mechanism of the droplet drying is similar for both small and large diameters. Therefore, it is recommended that a more accurate technique for measuring droplets with small sizes and for the taking of surface temperatures be developed.

The new model predicted temperature distribution profiles for single droplets of various materials. The predicted results showed a good agreement with the experimental data for air temperatures at 75° C and 140° C. However, the model was less accurate in the case of organic paste due to the higher thermal conductivity of the formed crust. A correction factor should be developed and taken into account for such materials. The model provides a relatively fast and efficient way to simulate drying behavior over a range of drying conditions. The model also represents a useful tool in the design and optimization of spray drying processes.

Moisture content profiles clearly showed the two stages of the drying process. In addition, the moisture profiles supported the conclusions that the crust forming in the falling rate period decreased both the moisture content and the drying rate.

Through the results of the experimental work and the theoretical model for a droplet drying, a significant conclusion was obtained. It has always been wrongly assumed in the literature that there is no temperature distribution within the droplet. This concept has been corrected by the current research. All experiments for the three materials used showed a clear difference between the core and the surface temperatures of the droplet during the drying process.

3.5. General conclusion

The characteristics of the boundary layer have a great effect on the local heat and mass transfer coefficients and temperature distributions throughout a flat bed surface. Droplet drying is an important subject in drying science, since it provides more details about the drying mechanisms in order to optimize spray dryer equipment. However, this part of the drying field is rarely considered in the literature. Further research in this area seems essential to obtain better understanding of drying theory.

Wind tunnel definitely is considered one of the best tools to investigate and to study the mechanisms of drying process. The most important variables in any drying process such as air flow, temperature and humidity are usually easy to be controlled inside the wind tunnel. Through a mathematical approach and an experimental work using a wind tunnel, we highlighted on the role of the boundary layer on the interface behavior and the drying mechanisms for various materials of a flat plate surface and a single droplet shape.

Nomenclature

α	Thermal diffusivity	m^2/hr
Cp	Heat capacity	kJ/kg K
D	Diameter	M
D_{eff}	Effective diffusivity	m^2/s
D_v	Diffusivity of vaporization	m^2/s
H	Heat transfer coefficient	$W/m^2\,K$
Δh_v	Latent heat of vaporization	kJ/kg
K	Thermal conductivity	W/m K
k'_c	Mass transfer coefficient	m/s
K'_G	Overall mass transfer coefficient	m/s
M	Molecular weight of water	kg/kg_{mol}
Nu	Nussult number	-
P	Partial pressure	kPa
Pr	Prandtel number-	

Q	Conduction heat transfer	W/m²
R	Universal constant	m³kPa/kg$_{mol}$ K
Re	Reynolds number	-
t	Time	s
T	Temperature	K
X	Moisture content	kg/kg
T$_a$	Air Temperature	K
ρ	Density	kg/m³
Sc	Schmidt number	

Subscripts

a	Dry crust
nz	Glass nozzle
sr	Surface
v	Vapor
w	Wet core

Author details

Abdulaziz Almubarak[*]

College of Technological Studies, Department of Chemical Engineering, Kuwait

References

[1] Sparrow E, Lin S. Boundary Layer With Prescribed Heat Flux Application to Simultaneous Convective and Radiation. International Journal of Heat and Mass Transfer 1965; 8, 437- 448.

[2] Luikov A. Conjugate Convective Heat Transfer Problems. International Journal of Heat and Mass Transfer 1974; 17, 257-265.

[3] Chyou T. The effect of A Short Unheated Length and A Concentrated Heat Source on The Heat Transfer Through A Turbulent Boundary Layer. International Journal of Heat and Mass Transfer 1991; 34, 1917-1928.

[4] Harris S, Ingham D, Pop I. Transient Boundary Layer Heat Transfer From A Flat Plate Subjected To A Sudden Change in Heat Flux. European Journal of Mechanics B- Fluid 2001; 20, 187-204.

[5] Deswita L, Nazar R, Ahmad R, Ishak A, Pop I. Similarity Solutions of Free Convection Boundary Layer Flow on a Horizontal Plate with Variable Wall Temperature. European Journal of Scientific Research 2009; 27(2) 188-198.

[6] Defraeyea T, Houvenaghelc G, Carmelieta J, Derome D. Numerical Analysis of Convective Drying of Gypsum Boards. International Journal of Heat and Mass Transfer 2012; 55, 4487-4928.

[7] Mori S, Nakagwa H, Tanimoto A, Sakakibara M. Heat and Mass Transfer With A Boundary Layer Flow Past A Flat Plate of Finite Thickness. International Journal of Heat and Mass Transfer 1991; 34, 2899-2909.

[8] Masmoudi W, Prat M. Heat and Mass Transfer Between A Porous Medium and A Parallel External Flow; Application to Drying of Capillary Porous Material. International Journal of Heat and Mass Transfer 1991; 34, 1975-1989.

[9] Jomaa W, Bruneau D, Nadeau J. Simulation of The High Temperature Drying of A Past Product: on The Influence of The Local Air Flow and The Thermal Radiation. Drying Technology 2004; 22, 1709-1729.

[10] Ranz W, Marshall W. Evaporation From Drops. Chemical Engineering Progress 1952; 48, 141-173.

[11] Trommelen A, Crosby E. Evaporation and Drying of Drops in Superheated Vapors. AIChE Journal 1970; 16, 857-872.

[12] Sirignano, W.A. Fluid Dynamics and Transport of Droplets and Sprays: Cambridge University Press, NY, USA 1999.

[13] Masters K. Spray Drying in Practice. Spray Dry Consult International: Denmark: John Wiley & Sons Inc., NY, USA 2002.

[14] Sloth J, Kiil S, Jensen A, Andersen S, Jørgensen K, Schiffter H, Lee G. Model Based Analysis of The Drying of A Single Solution Droplet in An ultrasonic Levitator. Chemical Engineering Science 2006; 61, 2701-2709.

[15] Miura K, Miura T, Ohtani S. Heat and Mass Transfer to and From Droplets. American Journal of Chemical Engineers, Symposium series 1977; 73 (163).

[16] Ali H, Mumford C, Jefferys G., Bains G. A study of evaporation from, and drying of, single droplets. In: Mujumdar AS. (editor) Proceedings of the 6th International Symposium in Drying IDS'88. Versailles, France.1988.

[17] Minoshima H, Matsushima K, Liang H, Shinohara K. Estimation of Diameter of Granule Prepared by Spray Drying of Slurry with Fast and Easy Evaporation. Journal of Chemical Engineering of Japan 2002; 35, 880-885.

[18] Seydel P. Experimental and Mathematical Modeling of Solid Formation at Spray Drying. Chemical Engineering Technology 2004; 27 (5): 505-510.

[19] Perdana J, Fox M, Schutyser M, Boom R. Single-Droplet Experimentation on Spray Drying: Evaporation of a Sessile Droplet. Chemical Engineering & Technology 2011; 34 (7): 1151–1158.

[20] Cheong H, Jeffreys G, Mumford C. A Receding Interface Model For The Drying of Slurry Droplets. AIChE Journal 1986; 32, 1334-1346.

[21] Farid M. A New Approach to Modeling of Single Droplet Drying. Chemical Engineering Science 2003; 58, 2985-2993.

[22] Almubarak A, Mumford C. Characteristics of the receding evaporation front in convective drying. In: Mujumdar AS. (editor) Proceedings of the 9th International Drying Symposium, IDS'94, Marcel Dekker Inc., NY, USA; 1994.

[23] Kays W. M, Crawford M.E. Convective Heat and Mass Transfer, 2nd. ed., McGraw-Hill Book Company, New York, NY, USA.1980.

[24] Thomas L. C. Heat Transfer professional Version. 2nd ed.: Capston Publishing Corporation, OK. USA. 1999.

[25] Kreith F, Bohn M. Principles of Heat Transfer. 6th ed.: Brook/Cole Publishers, Pacific Grove, CA, USA 2001.

[26] Almubarak A, Al-Saeedi J, Shoukry M. Effect of Boundary Layer on Mechanisms of Beach and Desert sand. European Journal of Soil Science 2008; 59, 807-816.

[27] Waananen K, Litchfield J, Okos M. Classification of Drying Models for Porous Solids, Drying Technology 1993; 11, 1-40.

[28] Holman J.P. Heat Transfer: McGraw-Hill Book Co., NY, USA 2002.

[29] Nesic S, Vodnik J. Kinetics of Droplet Evaporation. Chemical Engineering Science 1991; 46, 527-537.

[30] Chen X, Peng F. Modified Biot Number in The Context of Air Drying of Small Moist Porous Objects. Drying Technology 2005; 23, 83-103.

[31] Hayder M, Mumford C. 1993. Mechanisms of Drying of Skin Forming Materials. Drying Technology 1993; 11, 1713-1750.

Chapter 7

Wind Tunnel Tests on Horn-Shaped Membrane Roof Under the Turbulent Boundary Layer

Yuki Nagai, Akira Okada, Naoya Miyasato,
Masao Saitoh and Ryota Matsumoto

Additional information is available at the end of the chapter

1. Introduction

In this paper, the authors describe about a wind tunnel test for a membrane roof on a civil engineering. Especially, the authors focused on the horn-shaped membrane roof (shown in fig.1). Wind loading is the most dominant load for light-weight structures such as membrane roofs. A wind-force coefficient of typical building type such as box-type is defined in the guideline and the cord, but a wind-force coefficient of complicated shapes such as the horn-shaped membrane roof has not been sufficiently reported yet.

In general, there are two types of wind-tunnel test on the membrane roof, namely a test using a rigid model and a test using an elastic model. The test of the rigid model is used to measure the wind pressure around the building. On the other hand, the test of the elastic model can measure the deflection of the membrane surface directly and grasp the behavior of the membrane. This paper describes about the test using the rigid model for the horn-shaped membrane roof structure to measure a wind-force coefficient and fluctuating wind pressure coefficient around membrane under the turbulent boundary layer flow.

1.1. Past research about the wind tunnel on the horn-shaped membrane structures

Wind pressure coefficients of typical building type such as box-type are defined in guidelines and standards in each country, but wind pressure coefficients of complicated shapes such as the horn-shaped membrane roof have not been sufficiently reported yet.

The basic studies, which were about the theory and the analysis method, on the horn-shaped membrane roof were reported by F. Otto, M. Saitoh et al and also were shown the wind-pressure coefficients of the horn-shaped membrane roof under regulated conditions in several

Stand-alone Model	Multi-bay Model
Rest Dome (1989)	Kashiwa no Mori (2008)
Tsukuba Expo., Japan (1985)	Hyper Dome E (1990)

Figure 1. Horn-shaped membrane roof

reports and books (Otto, 1969; Saitoh & Kuroki, 1989; Nerdinger, 2005). In the recent years, studies on the numerical simulation against the horn-shaped membrane roof were reported by J. Ma, C. Wang et al (Ma et al., 2007; Wang et al., 2007). Furthermore, dissertation by U. Kaiser indicated wind effects on weak pre-stressed membrane structure which is 30m horn shaped membrane by aero-elastic models (Kaiser, 2004).

In this way, there are many other references on this field. However, the basic date for the wind-force coefficient of the horn-shaped membrane roof has not been sufficiently reported yet. Based on this background, the authors have carried on the wind tunnel test, and report these results.

1.2. The composition of this paper

In this chapter, the authors describe about a composition of this paper and explain three types of wind tunnel test (see fig. 2).

Chapter 2 shows a form of the horn-shaped membrane roof and example of a basic technique to form finding method of the membrane structure before the wind tunnel tests. Chapter 3 shows definitions of symbols and calculation formulas on this paper. Chapter 4 shows outline of wind tunnel device and method of measuring. Chapter 5 shows a flow condition of the test which is the turbulent boundary layer flow, and test conditions. Chapter 6 and 7 show the wind tunnel tests and the results; the test of stand-alone type model in chapter 6 and the test of multi-bay models on chapter 7. These tests indicate mean wind pressures coefficient, fluctuating wind pressure coefficient and peak wind pressure coefficient around the horn-shaped membrane structures under the turbulent boundary layer flow.

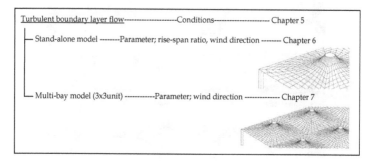

Figure 2. The composition of this paper

2. Form of the horn-shaped membrane roof

The horn-shaped membrane roofs have several kind of planar shape, namely a circle, a square and a hexagon. This paper describes about the square based horn-shaped membrane roof. In general, the membrane structure needs to find appropriate forms to resist external force. 'European Design Guide for Tensile Surface' by TensiNet presents some methods of form-finding for the membrane structures (Forster & Mollaert, 2004). This paper used nonlinear finite element method to find the appropriate form on the square based horn-shaped membrane.

In this paper, the membrane material was defined as low stiffness material (see figure 3). On the other hand, a strut was defined as high stiffness material. A strut was transferred point B from point A in order to get the appropriate form using FEM analysis. A rise-span ratio h/L was defined as the ratio of a span L to a height of the horn-shaped roof H, and an appropriate form of h/L=0.2 was obtained by finite element method with geometrical nonlinear in this paper. Additionally, the top of strut was L/10 and there wasn't a hole on the middle of the horn-shaped roof. The final shape get three-dimensional curved surface.

3. Definitions of symbols and calculation formula on this paper

The wind pressure coefficient was calculated based on *The Building Standard Law of Japan* (The building Center of Japan, 2004), *Recommendations for Load on Buildings 2004* (Architectural Institute of Japan, 2004) and *ASCE Manuals* (Cermak & Isyumov, 1998). Definitions of the symbols in this paper are shown in figure 4. As for the signs of wind pressure coefficient, the positive (+) means positive pressure against the roof and the negative (-) means negative pressure against the roof.

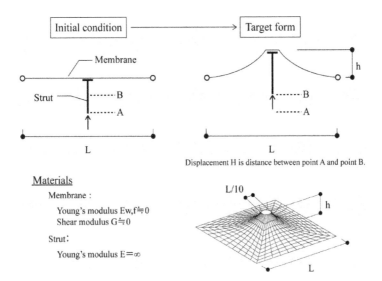

Figure 3. Form finding method on the horn-shaped membrane structure

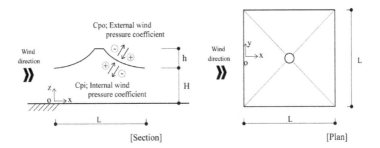

Figure 4. The definitions of symbols in this paper

The wind pressure coefficient is obtained from follows;

$$C_{pj} = C_{poj} - C_{pij} \tag{1}$$

$$C_{pij} = \frac{P_{ij} - P_s}{\bar{q}_z}, C_{poj} = \frac{P_{oj} - P_s}{\bar{q}_z} \tag{2}$$

$$\bar{q}_z = \frac{1}{2}\rho \bar{v}_z^2 \tag{3}$$

in which C_{pj} is the wind pressure coefficient at measurement pressure tap j, C_{poj} is the external wind pressure coefficient at measurement tap j, C_{pij} is the internal wind pressure coefficient at measurement tap j, P_{ij} is the internal pressure at measurement tap j, P_o is the external pressure at measurement tap j, P_s is the static, or the barometric, pressure at a reference location, $\bar{q}_z$ is the mean value of dynamic pressure at the reference location z, ρ is the density of the air, and $\bar{v}_z$ is the mean value of wind velocity at the reference location z. In this paper, the reference location z with the uniform flow means the position of the pitot tube. On the other hand, the reference location z with the turbulent boundary layer flow was obtained from the following equations;

$$z = h + \frac{H}{2} \tag{4}$$

in which h is the eave height of the roof, and H is the rise of the horn-shaped roof.

Particularly, the mean value of wind pressure coefficient C_{p_mean} and the peak value of wind pressure coefficient C_{p_peak} are expressed respectively as follows;

$$C_{p\_mean} = C_{po\_mean} - C_{pi\_mean} \tag{5}$$

$$\begin{cases} C_{p\_peak,\max} = C_{po\_peak,\max} - C_{pi\_peak,\min} \\ C_{p\_peak,\min} = C_{po\_peak,\min} - C_{pi\_peak,\max} \end{cases} \tag{6}$$

in which C_{po_mean} and C_{pi_mean} are the mean value of external and internal wind pressure coefficient, C_{po_peak} and C_{pi_peak} are the tip value of external and internal wind pressure coefficient.

Additionally, C_{pi_mean}, C_{po_mean}, C_{po_peak} and C_{pi_peak} are given by the following equations;

$$C_{pi\_peak} = \frac{P_{i\_mean}}{\bar{q}z}, C_{po\_peak} = \frac{P_{o\_mean}}{\bar{q}z} \tag{7}$$

$$C_{pi\_mean} = \frac{P_{i\_mean}}{\bar{q}z}, C_{po\_mean} = \frac{P_{o\_mean}}{\bar{q}z} \tag{8}$$

in which P_{i_mean} and P_{o_mean} are the mean value of internal and external wind pressure on the pressure measurement tap respectively, and P_{i_peak} and P_{o_peak} are the tip value of internal and external wind pressure on the tap. In case of the enclosed type which is constructed with side walls, P_i is neglected on these calculations.

4. Outline of wind tunnel configuration

These tests were aimed at measuring local wind pressure on the horn-shaped membrane roof using the Eiffel type wind tunnel as shown in table 1 and figure 5. The turbulent boundary layer flow was made by the roughness blocks, the spires and the trips (show in figure 6). The P_j-P_s, which P_j is the pressure at the measurement pressure tap j and P_s is the static pressure at the pitot tube, was measured directly by the laboratory pressure transducer as a differential pressure and represents the wind pressure acting at the particular pressure tap location j within the computer as sown in figure 7.

	Wind tunnel facility	Eiffel type wind tunnel
	Length of wind tunnel	31000mm
Wind tunnel	Section size	2200×1800×17300mm (width×height×length)
	Contraction ratio	1 : 3
	Velocity range	0.0~25.0
	Form	GFPR's axial fan
Blower	Wing shape	φ=2500mm
	Volume	About 100

Table 1. Outline of wind tunnel configuration

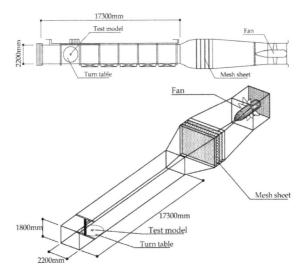

Figure 5. Sketch of Eiffel wind tunnel used

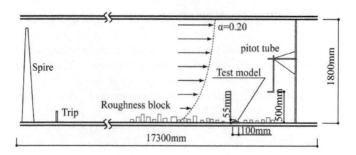

Figure 6. Cross-section diagram of wind tunnel facilities

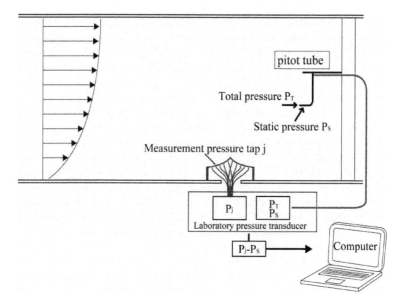

Figure 7. The wind pressure acting at the particular pressure tap location j

5. Outline of the turbulent boundary layer flow

In this chapter, the outline of the turbulent boundary layer flow is described. Table 2 shows conditions and parameters on the tests. It was assumed that a model scale was 1: 100 and that a velocity scale was 7/27 at the full scale wind speed 34m/s. In this case, time scale was 11/125, and additional flow conditions indicate in figure 9. Airflow conditions which were the average wind speed profile, the turbulence intensity, the power spectral density of fluctuating wind speed and the scale of turbulence for this test, are shown in figure 9. The velocity gradient α was 0.2 and the turbulent intensity around the roof was about 0.3. This wind was simulated natural wind in the urban area, namely "terrain 3" in the Building Standard Low of Japan.

Figure 8. Photos of wind tunnel test

Flow	Boundary Turbulent Layer Flow (Urban Area; Terrain 3 in The Building Standard Law of Japan)
Wind velocity	About 7 m/s at z=35mm (around the test model)
Velocity gradient α	α=0.2
Velocity turblence intensity Ir	0.3 at z=35mm (around the test model)

Table 2. Airflow Condition on the wind tunnel

Model Type	Stand-alone model, Multi-bay Model
Sampling speed	500Hz
Sampling time	30sec
Rise-span ratio h/L	0.1, 0.2, 0.3
Model scale	100mm x100mm (model : full =1:100)
Wall	Open type / Enclosed type
Wind direction	0-degree, 15-degree, 30-degree, 45-degree
Number of test on each model	Five times

Table 3. Model Condition

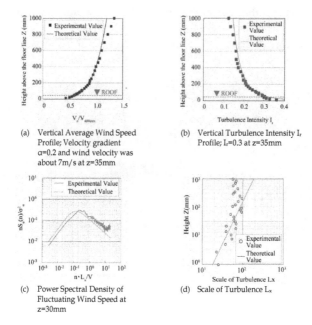

(a) Vertical Average Wind Speed Profile; Velocity gradient α=0.2 and wind velocity was about 7m/s at z=35mm

(b) Vertical Turbulence Intensity I_r Profile; I_r=0.3 at z=35mm

(c) Power Spectral Density of Fluctuating Wind Speed at z=30mm

(d) Scale of Turbulence L_x

Figure 9. Wind flow conditions in the wind tunnel test

6. The wind tunnel test on the stand-alone model under the turbulent boundary layer flow

This chapter focuses on the stand-alone model of horn-shaped membrane roof and indicates wind pressure and fluctuating pressure around models under the boundary turbulent layer flow which was shown in the preceding section.

6.1. Outline of tests

The 100mm x 100mm square based model was used in this test. Major parameters were three types of rise-span ratio (h/L), namely h/L=0.1, 0.2 and 0.3, and the presence of walls. Six types of model were prepared for this wind tunnel test. The outline of models and measurement taps show in figure 10 and figure 11.

These models were made from acrylic plastic. As for the open type model, the roof depth was about 5mm in order to measure both sides of the roof at the same time (show in figure 12). Additionally, wind directions were only four types which were 0-deg., 15-deg., 30-deg. and 45-deg., because of symmetry form of roof.

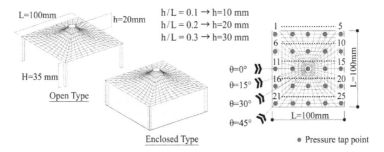

Figure 10. Experimental models and measuring points on the stand-alone models; two types model was prepared, namely "Open type" and "Enclosed type"

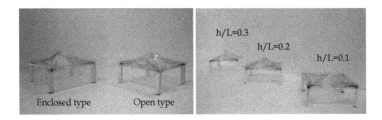

Figure 11. The photo of models; three types of h/L models which was made from acrylic plastic. The depth of open type's roof is about 5mm thick.

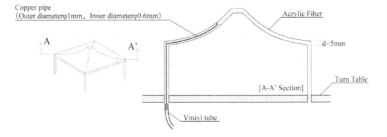

Figure 12. Details of the experimental model

6.2. Results of mean wind pressure coefficient on the stand-alone model

Distributions of mean wind pressure coefficient on each model are indicated in figure 13 and 14. The distribution of wind pressure coefficient changed the value depending on the presence of the wall. Similarly, the wind pressure coefficient distributions depended on the wind direction.

In the open type, the negative pressure concentrated at the windward side on the model. On the other hand, the negative pressure observed at the top of the roof on the enclosed model. Moreover, the negative pressure around the top of roof was increase with increasing of a rise-span ratio.

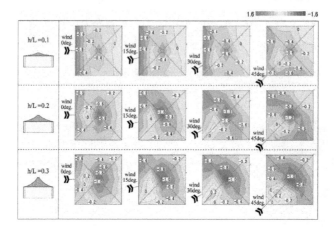

Figure 13. Mean wind pressure coefficient which was obtained from wind tunnel tests on enclosed type of the stand-alone mode

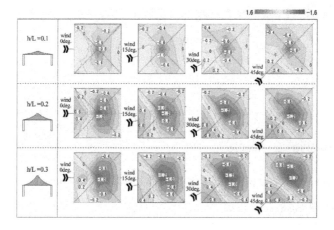

Figure 14. Mean wind pressure coefficient which was obtained from wind tunnel tests on open type of the stand-alone mode

6.3. Results of fluctuating wind pressure coefficient on the stand-alone model

This section shows the distributions of fluctuating wind pressure coefficient on each model (show in figure 15 and 16). The fluctuating wind pressure coefficient C_f' was obtained from the following equations;

$$C_f' = \frac{\sigma_p}{\bar{q}_z} \tag{9}$$

in which σ_p is fluctuating wind pressure at pressure tap p on the model and $\bar{q}_z$ is the mean value of dynamic velocity pressure at the reference location. The maximum value of the fluctuating wind pressure is "1.0" and the minimum value of the fluctuating wind pressure is "0".

The test result showed that the C_f' of the enclosed types were different distribution from the open types. Furthermore the C_f' of the enclosed type was larger than that of the open type. Especially, the model type h/L=0.2 of the enclosed model showed 0.75 around the center of the roof. These results may cause some effects on the response of membrane, since the membrane structure is generally sensitive structure for the external force such as wind load with turbulence.

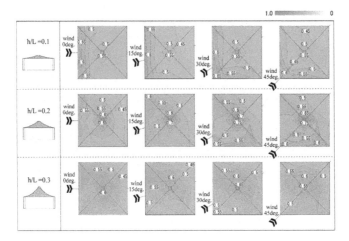

Figure 15. Fluctuating wind pressure coefficient which was obtained from wind tunnel tests on enclosed type of the stand-alone mode

6.4. Results of peak wind pressure coefficient on the stand-alone model

Distributions of the peak wind pressure coefficient on each model are indicated in figure 17 and 18. Generally, the peak wind pressures around corner of roof distinct from distributions

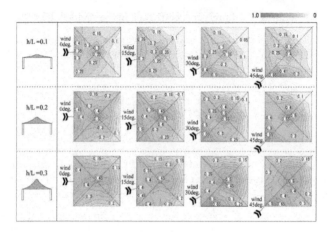

Figure 16. Fluctuating wind pressure coefficient which was obtained from wind tunnel tests on open type of the stand-alone mode

of the internal area. However, this test showed that peak wind pressure coefficients around the middle of roof (i.e. the top of roof) were the maximum negative value. In addition, the peak wind pressure coefficient of the enclosed model was larger than that of the open type. For example, focusing on the enclosed model, the model of h/L=0.2 and 0.3 show more than -4.0. Furthermore, the distribution varied according to the parameter of wind direction and rise-span ratio.

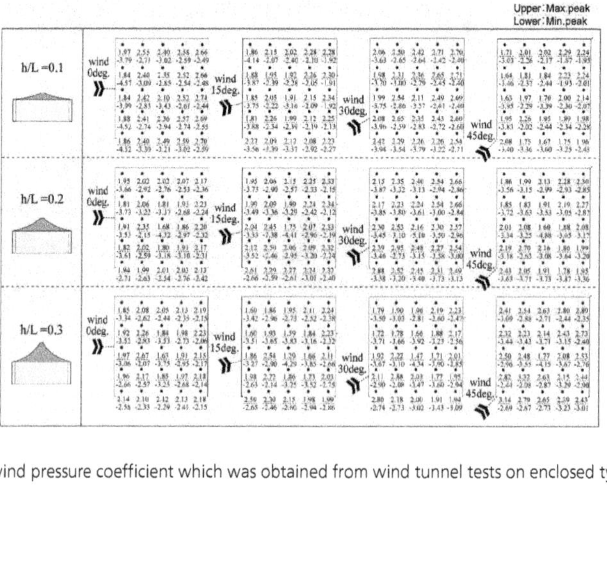

Figure 17. Peak wind pressure coefficient which was obtained from wind tunnel tests on enclosed type of the stand-alone mode

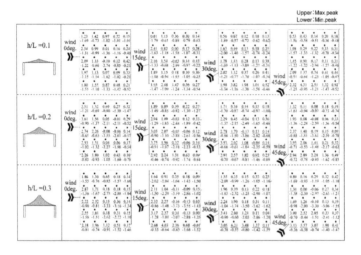

Figure 18. Peak wind pressure coefficient which was obtained from wind tunnel tests on open type of the stand-alone mode

7. The wind tunnel test on the multi-bay model under the turbulent boundary layer flow

In most cases, the horn shaped membrane structure is used as the multi-bay type. The number of horn unit depends on the scale of the building and the building uses. Therefore, this chapter focuses on the multi-bay model of 3×3. This test was carried out to clarify about the basic characteristics of the wind pressure coefficient of the multi-bay horn-shaped membrane roof.

7.1. Outline of tests

This test used the same facilities and the same turbulent flow as the stand-alone model shown in chapter 5. A model scale of a horn unit was 30cm x 30cm and the number of unit was 3 wide, 3 bays, and the models ware made from acrylic (see figure 19 and 20). This experimental model was only one type of rise-span ratio, namely h/L=0.2.

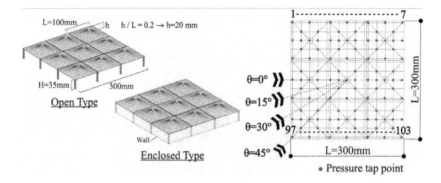

Figure 19. Experimental models and measuring points on the multi-bay models

Figure 20. The photo of models on the multi-bay model; one type of h/L model which was made from acrylic plastic.

7.2. Results of mean wind pressure coefficient on the multi-bay model

Distributions of mean wind pressure coefficient on each model are shown in figure 21 and 22. The distributions were changed by wind direction as same as stand-alone models. Focusing on the enclosed model, the positive pressure were shown around the valley of the roof. On the other hand, in the open type, windward side show positive pressure.

These results of open type were obtained approximately the same results with the stand-alone model of open type. On the other hand, as for the enclosed type, results were different from the stand-alone model. Specifically, focusing on the rise-span ratio 0.2, the value of the wind pressure coefficient around the middle of model was smaller than the stand alone models.

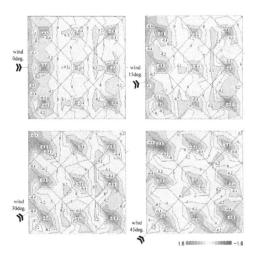

Figure 21. Mean wind pressure coefficient which were obtained from wind tunnel tests on enclosed type of the multi-bay mode

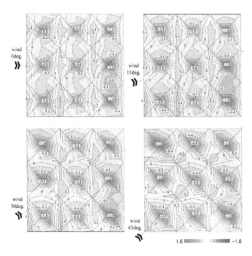

Figure 22. Mean wind pressure coefficient which were obtained from wind tunnel tests on open type of the multi-bay mode

7.3. Results of fluctuating wind pressure coefficient on the multi-bay model

Distributions of fluctuating wind pressure coefficient on each model are indicated in figure 23 and 24. The fluctuating wind pressure coefficients indicated on multi-bay model almost the same as that on stand-alone model. The enclosed model showed value of 0.6 or more over the whole area of the roof. But the open type showed comparatively large value of approximately 0.8 on the only windward side.

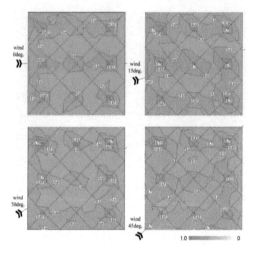

Figure 23. Fluctuating wind pressure coefficient which were obtained from wind tunnel tests on enclosed type of the multi-bay mode

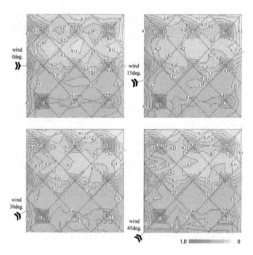

Figure 24. Fluctuating wind pressure coefficient which were obtained from wind tunnel tests on open type of the multi-bay mode

7.4. Results of peak wind pressure coefficient on the multi-bay model

The maximum peak wind pressure coefficients are shown in figure 25, and the minimum peak wind pressure coefficients are shown in figure 26. These distributions were changed by wind direction. Furthermore, these wind pressure coefficients around the top of roof indicated the maximum negative value. And these results were smaller than the stand-alone models.

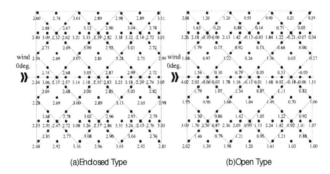

Figure 25. Maximum peak wind pressure coefficient which were obtained from wind tunnel tests on the multi-bay model

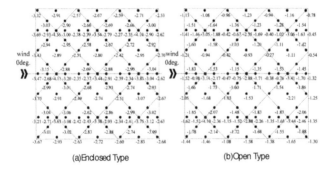

Figure 26. Minimum peak wind pressure coefficient which were obtained from wind tunnel tests on the multi-bay model

8. Conclusions

In this paper, the characteristics of the wind pressure coefficients on the horn-shaped membrane roof were presented using wind tunnel tests with the turbulent boundary layer flow. Particularly, the followings are clarified that;

- The wind pressure coefficient varied according to the presence of the wall and the wind direction.

- The negative pressure around the top of roof become larger with the increase of the rise-span ratio.

- The fluctuating wind pressure coefficient and the peak wind pressure coefficient on the enclosed type was larger than these of the open type.

- As for the mean wind pressure coefficient of the enclosed types, the multi-bay types were different from the stand-alone models. These results are forecast to cause unstable phenomenon of the membrane.

Furthermore, the representative distributions of the wind pressure coefficient were shown on each parameter.

Acknowledgements

This study was supported by Japan Society for the Promotion of Science, Grant-in-Aid for JSPS Fellows, KAKENHI 22·7895. All of tests were carried out on "Research Institute of Science and Technology, College of Science and Technology, Nihon University". The authors have had the support of Ayu Matsuda, Japan ERI Co.,Ltd., and Tomoaki Kaseya Graduate School of Science and Technology, Nihon University to carry out the experiments.

Author details

Yuki Nagai[1], Akira Okada[2], Naoya Miyasato[2], Masao Saitoh[2] and Ryota Matsumoto[2]

1 Sasaki Structural Consultants, Japan

2 Nihon University, Japan

References

[1] Architectural Institute of Japan, *Recommendations for Load on Buildings. (2004)*, Architectural Institute of Japan, ISBN 481890556,Japan

[2] The building Center of Japan. (2004). *The Building Standard Law of Japan June 2004*, The building Center of Japan. , ISBN 4-88910-128-4, Japan

[3]   Cermak, J.E. & Isyumov, N., with American Society of Civil Engineers Task Commit-tee. (1998), *Wind Tunnel Studies of Buildings and Structures (Asce Manual and Reports on Engineering Practice)*, American Society of Civil Engineers, ISBN 0784403198

[4]   Cook, N.J. (1990), *Designer's Guide to Wind Loading of Building Structures Part 2: Static structures*, Laxton's, ISBN 0408008717

[5]   Forster, B. et al. (2004), *European Design Guide Tensile Surface Structures*, TensiNet, ISBN 908086871

[6]   Kaiser, U. (2004), *Wind Wirkung auf Schwach Vorgespannte membran strukturen am bei-spiel eines 30m-membranschirmes*, Der Andere Verlag., ISBN 3899591623, Germany

[7]   Ma, J., Zhou, D., LI, H., ZHU, Z. & DONG, S. *Numerical simulation and visualization of wind field and wind load on space structure*, Proceedings of IASS 2007, Beijing, 2007

[8]   Nerdinger, W. (2005). *Frei Otto Complete Works: Lightweight Construction Natural De-sign*, Birkhäuser Architecture, ISBN 3764372311

[9]   Janberg, N. (2011). BC Place stadium, In: *Nicolas Janberg's Structurae,* March 21, 2011, Available from: http://en.structurae.de/structures/data/index.cfm?id=s0000708

[10] Janberg, N. (2011). Lord's Cricket Ground Mound Stand, In: *Nicolas Janberg's Struc-turae,* March 21, 2011, Available from: http://en.structurae.de/structures/data/index.cfm?id=s0000694

[11] Otto, F. (1969). *Tensile Structures: Cables, Nets and Membranes v. 2*, MIT Presse, ISBN 0262150085, USA

[12] Saitoh, M. (2003). *Story of Space and Structure -Structural Design's Future*, Shoukoku-sha, ISBN 4395006396, Japan

[13] Saitoh, M. & Kuroki, F. *Horn Type Tension Membrane Structures*, Proceedings of IASS 1989, Madrid, 1989

[14] Seidel, M. & David, S. (2009). *Tensile Surface Structures - A Practical Guide to Cable and Membrane Construction: Materials, Design, Assembly and Erection*, Wiley VCH, ISBN 3433029229, Germany

[15] Shinkenchiku-Sha Co. Ltd. (1992). Hyper Dome E, In: *Shinkenchiku March,1992*, Shin-kenchiku-Sha Co. Ltd. ISSN 1342-5447, Japan

[16] Shinkenchiku-Sha Co. Ltd. (1988). Tokyo Dome, In: *Shinkenchiku May, 1988*, Shinken-chiku-Sha Co. Ltd. ISSN 1342-5447, Japan

[17] Shinkenchiku-Sha Co. Ltd. (2007). BDS Kashiwanomori Auctionhouse, In: *Shinkenchi-ku October, 2007*, Shinkenchiku-Sha Co. Ltd. ISSN 1342-5447, Japan

[18] Wang, C., Zhou, D. & Ma, J. *The interacting simulation of wind and membrane structures*, Proceedings of IASS 2007, Beijing, 2007

Experimental Study of Internal Flow Noise Measurement by Use of a Suction Type Low Noise Wind Tunnel

Yoshifumi Yokoi

Additional information is available at the end of the chapter

1. Introduction

The measurement experiment of the fluid-dynamic noise made from the object placed into the air flow is performed using a low noise wind tunnel, a silent airflow wind tunnel, etc. In the low noise wind tunnel, the measures against silence are taken so that the noise generated with a fan or a compressor may not propagate as much as possible to a wind tunnel test section by an air current. As for the surroundings of the test section of a low noise wind tunnel, acoustic free space is provided. Generally a wind tunnel is classified by the form of the channel of a wind tunnel (blow type, suction type and circulating type), the form of the measurement section (open, half-open and sealed), and the existence of circulation of flow. And the practical wind tunnels are classified into 13 kinds (Mochizuki & Maruta, 1996). Figure 1 illustrates the circulation environment for the airflow between the blower and the measurement section, the types of duct in the wind tunnel (blow, suction and circulating) and the types of measurement section (open, half-open and sealed). The merit of each type of the wind tunnel and the weak point are summarized as follows. In the merit of the blow type, the composition is simple and small the installation space. In the liberating measurement section of jet-type, the usage of use becomes various. The week point is to need big power because the pressure loss is large. Flowing quantity will come to receive the fluctuation easily in turbulence. The measurement section is that the temperature raises more than the temperatures of air in the surrounding. The merit of the suction-type should be able to be composed the rectification part short, and more compactly than the blow-type. The temperature of the measurement section is the same as the temperature of the space in the surrounding. The weak point is to receive the influence of the fluctuation of the outer air flow large. The measurement section must become negative pressure from the atmospheric pressure. An enough space for the rectification is needed on the suction side. The merit of the circulating blow-type is not to receive turbulence. The experiment on all-round is possible in the open-type measurement section. The weak point is to take time until stabilizing and it be easy to rise in the temperature. Merits of the circulating suction-type are that turbulence is not received and the rectification part is short. The weak point is to take time until stabilizing. A very wide space is necessary forward of the suction mouth. Merits of the circulating type to unnecessary big power and not to receive turbulence. The stability of the flow is also early. Especially, efficiency is very good and the pressure loss is a little in the sealed-type measurement section. The weak point is to need noting in the rise's of the air flow temperature becoming remarkable. The object flow must be limited. A wide installation space is needed. In addition, there are a peculiar merit and a weak point respectively by the measurement section shape, and they are summarized as follows. The merit of the open-type measurement section is that the limitation concerning the size and the shape of the test piece is a little. The weak point is to receive turbulence by the suck of air. Merits of the half-open type measurement section are permitted the test piece diversity and are hard of turbulence to receive. The weak point is that the measurement room becomes negative pressure easily. The merit of the sealed-type measurement section is to become the most efficient wind tunnel, and to hardly receive turbulence. The weak point is to receive the limitation to the size and the shape of the test piece. Among these, it is required that the wind tunnel aiming at measurement of a fluid-dynamic noise secures the acoustic free space of silence and a test section. Moreover, it is also required that the spatial relationship of a test model and a microphone can be set up freely. Therefore, many blow-type wind tunnels with the measurement room and half-open type test section by which sound insulation processing was carried out with the sound-absorbing material are used. On the other hand, use of a microphone is difficult in an air flow, and the measurement technique of a fluid-dynamic noise has not been established. Therefore, the wind tunnel with a sealed type test section can scarcely be seen. Accordingly, measurement of the fluid-dynamic noise of internal flows, such as a flow inside a gas turbine or a jet engine, and a pipeline, a flow of the around of the support in a duct, is not in the state which can be performed immediately. As for the present condition, there are also few examples of verification of measurement of the fluid-dynamic noise of an internal flow. So, it is very important to establish the measurement technique of the fluid-dynamic noise of an internal flow in engineering. In measurement of the fluid-dynamic noise using a low noise wind tunnel, when an open-type test section is used, it is reported that there is a case where it becomes impossible for a back ground noise not to be amplified by the large turbulence produced with the edge of the jet stream from a nozzle, or for generating of the sound which is not a measuring object to be observed by interference of a jet and a model sample, or to maintain the two dimensional characteristic of a flow etc. Moreover, when a sealed type test section is used, on the usual surface of a wall, sound reflects, and exact measurement cannot be performed, but if the material which can bear wind pressure that sound tends to penetrate the surface of a wall is used, it is reported that the sealed type test section will probably be better (Fujita, 1994, 1996).

The purpose of this study is examination of the measurement technique of the fluid-dynamic noise of an internal flow. In this study, it proposes carrying out burial setting of the microphone to the test section equipped with a fibered glass. The suction type low noise wind tunnel with such a test section for verification was created, and measurement of the fluid-dynamic noise made from the circular cylinder placed into the air flow was tried. Comparison examination of the measurement result obtained by this measurement technique was carried out with the measurement result obtained in the blow type wind tunnel. As a result, it was shown that the same characteristic is obtained about the change in a sound pressure level or peak frequency. Moreover, since the target acoustic frequency was caught clearly, it was shown that it is convenient for examination of an acoustic effect. This measurement technique showed clearly that usefulness is high to fluid-dynamic noise measurement of the internal flow.

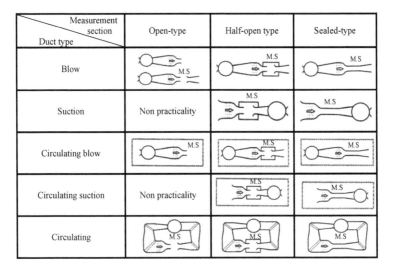

Figure 1. Wind tunnel classifications (Mochizuki & Maruta, 1996); the circle represents the blower, the arrow shows direction of the flow, and "M.S" is the measurement section

2. Experimental apparatus and method

This chapter describes the used equipment, a tool, and the procedure of an experiment.

2.1. Outline of the experimental apparatus

The experimental apparatus consists of a low noise wind tunnel and measuring equipment. Figure 2 shows the schematic diagram of a low noise wind tunnel. The low noise wind tunnel is constituted from the bell mouse, the test section, the silence duct, and the fan by the inhaled type wind tunnel with a sealed type test section. In order to reduce fan generating noise, the inside of a silence duct is divided into four in the shape of a cell, the sound-absorbing material (fibered glass) is stuck on all the surface of a wall, and the fan is installed in the fan room by which interior was carried out with the sound-absorbing material with a silence exhaust port with which three splitter walls were set. Regulation of airflow velocity which passes a test section is performed by carrying out inverter control of the number of rotations of the fan by remote control. A measuring device is divided roughly into fluid-dynamic noise measurement equipment and the air flow velocity measurement equipment. Fluid-dynamic noise measurement equipment consists of directive capacitor microphone (RION, UC-30, hereafter it is called microphone for convenient), precision noise level meter (RION, NA-40), and FFT analyzers (Ono Sokki, CF-5220). The air flow velocity measurement equipment consists of a hot-wire anemometer (DISA, TYPE55) and a digital pressure gauge (Cosmo Instruments, DM-3100B). As for measurement of the turbulence intensity to the flow velocity distribution and a main flow, the hot-wire anemometer was used. The pressure difference between the surface of a wall (static pressure) of a test section and atmospheric pressure was measured with the digital pressure gauge.

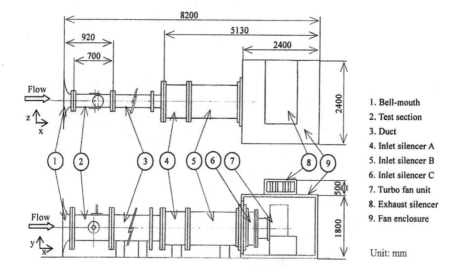

Figure 2. The schematic diagram of the wind tunnel

2.2. Measurement section and test cylinders

Figure 3 shows the schematic diagram of a measurement section (test section). The measurement section is a rectangular cross-section, 376mm (y direction) in height and 160mm (z direction) in width, with both side walls made of a transparent acrylic resin 700mm (x direction) in length, and a board thickness of 10mm. The turntable installation hole with a diameter of 100mm was installed from the edge of the measurement section upstream side to the position at 350mm in the centerline. Upper and lower walls act as the sound absorbing walls (fibered glass walls), with 50mm-thick fibered glass placed on a 15mm-thick transparent acrylic board. Half free space is made in acoustics by installing this sound absorbing wall. The microphone and the hot-wire probe are set up from the edge of the measurement part upstream side to the position at 400mm in the centerline. The surface of microphone and the surface of fibered glass are set at the same level. The hot-wire probe can be moved in a vertical direction in the measurement section (y direction) using the traverse device. The test circular cylinder can be set within a range of 5mm-45mm up from the center of the turntable. Here, the center-to-center spacing of the microphone and the circular cylinder make adjustments within a range of 5mm-95mm possible. The test circular cylinder is made from brass, span length is 160mm and the surface is finished smoothly. The test circular cylinder is with seven kind, and each diameter is 6mm, 10mm, 15mm, 20mm, 25mm, 30mm, and 40mm.

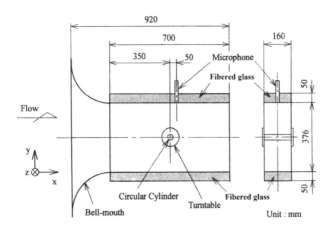

Figure 3. The schematic diagram of the test section (measurement section)

2.3. Experimental method and procedure

In advance of measurement of the fluid-dynamic noise, the flow velocity distribution in the test section is measured by a hot-wire anemometer, and the state of flow is understood. The relationship between the air flow velocity which passes the test section and the static pressure on the surface of wall is previously authorized using a Pitot tube and a digital pressure gauge. Proofreading of a microphone and a precision noise level meter is performed using the piston phone (RION, NC-72, 250Hz, 114dB). The measurement procedure for the sound of flow is as follows. The test air flow velocity is set by operating the rotational speed controller of the blower. The fluid-dynamic noise is measured by the microphone, and the overall noise level and frequency analyses are done using the precision sound level meter and the fast Fourier transform analyzer. The flow velocity distribution in the measurement section and the measurement of the disturbance intensity relative to the main flow is as follows. The I type probe of the hot-wire anemometer is inserted detaching the microphone, it traverses in a vertical direction (y direction) at 5mm intervals (the interval of traverse is 2.5mm near the wall), and the air flow velocity is measured at the microphone installation position. The frequency of the oscillating flow due to Karman vortex shedding from the circular cylinder is measured as follows. The I type probe of the hot-wire anemometer is fixed in a position such that a clear shape of the waves can be obtained, and the output signal and frequency are using the fast Fourier transform analyzer. Here, averaging is performed ten times in the frequency analysis.

3. Experimental result and discussion

This chapter describes the result of having investigated about the basic characteristic of a producing wind tunnel, and the result of having performed sound verification.

3.1. The fluid-dynamic characteristic and the acoustic characteristic of a producing wind tunnel

In order to understand the performance of a producing wind tunnel, investigation of the minimum flow velocity and the maximum flow velocity was performed using the Pitot tube. The minimum flow velocity in the test section was 2.5m/s, when the number of rotations of a fan was 100min^{-1}, and the maximum flow velocity in the test section was 35m/s when the number of rotations of a fan was 1300min^{-1}.

In a low noise wind tunnel, it becomes important especially to suppress propagation of the operation noise of the fan. Since this wind tunnel is a suction type wind tunnel, it is necessary to make it not accept fan generating noise in a test section. Accordingly, it is important not to leak the operation sound of the fan outside a fan room. So, the noise characteristic of the around of a wind tunnel was investigated. In order to understand the quietness of the wind tunnel, the sound pressure level around the test wind tunnel when it is driven or stopped was measured. Generally, the noise when the wind tunnel is operated is divided into air flow noise, and the operating noise of the blower. It is especially important in the fluid-dynamic noise measurement to suppress the propagation of the operating noise of the blower. The wind tunnel should not accept the blower generation noise in the measurement section. It is important that the operating sound of the blower does not leak outside the fan room. It is necessary, therefore, to understand the noise characteristics around the wind tunnel. The microphone positions for the noise measurement around the wind tunnel are shown in Fig. 4. Microphones are set up outside the fan room at a height of 1m off the ground, at measurement points (A-K). At measurement points (L1, L2) in the blower room, microphones are set up at a height of 1m, and placed a 700mm away from the electric motor and the blower outlet. Figure 5 shows the noise measurements at each measurement point when the circular cylinder is not set up in the measurement section and when the wind tunnel is in operation. The noise levels around the wind tunnel, almost the same, but differ inside and outside of the fan room, and when flow velocity increases, the difference increased. The noise levels inside and outside the fan room were 26dB and 32dB, respectively, when the wind tunnel was not operating. The level of sound intensity is defined by $L=10\log_{10}I/I_0$ (dB) (I_0 is an intensity of the sound of the standard: 10^{-12} W/m²). Here, when the level of intensity of a sound inside the fan room is defined as L_{IN}, and the level of intensity of a sound outside the fan room is defined as L_{OUT}, the ratio of the level of sound-intensity L_{IN}/L_{OUT} is given by $L_{IN}/L_{OUT}=10^{(LIN-LOUT)/10}$. The air flow velocity range is 5-35m/s, so the ratio of the level of sound-intensity L_{IN}/L_{OUT} becomes 155 -6760. Therefore, it is clear that the noise in the blower room is intercepted.

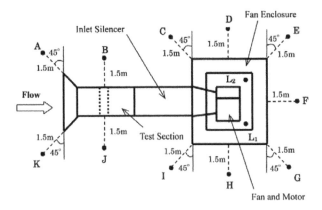

Figure 4. Sound measurement points around a wind tunnel pressure level

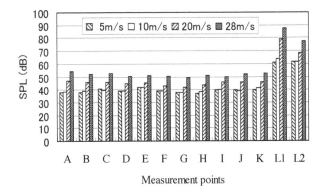

Figure 5. Sound pressure level around a wind tunnel with a flow velocity in the test section ranging from 5m/s to 28m/s

3.2. Flow characteristics in the measurement section

Flow characteristics in the measurement section where the sound absorbing wall (fibered glass wall) had been used were investigated. The hot-wire probe was inserted in the microphone's installation position; it traversed in a vertical direction (y direction), and the velocity and disturbance intensity were measured. Figure 6 shows the velocity distribution and the disturbance intensity when the air flow velocity is U=28m/s. The velocity distribution ($\triangle$ symbol) for an acrylic wall is plotted for comparison. Here, the abscissa is a measurement position. The center of width in the vertical direction of the measurement section is assumed to be zero points; the upper side is assumed to be + mark, and the lower side is assumed to be - mark. The measurement position is made dimensionless by width B=376mm in the vertical direction of the measurement section. The ordinate shows the velocity distribution and the disturbance intensity, respectively. The symmetry of a flow was good to the center and turbulence intensity was less than 0.5% in the range which is maintaining the equality of a flow. The disturbance intensity to keep the uniformity of the flow was within 0.5%. Here, if uniform flow velocity U=28m/s and the measurement point x=400mm are calculated by Prantl's exact solution (δ=0.22$(v/Ux)^{0.167}x$), the thickness of the boundary layer is 9.6mm. The thickness of the boundary layer as in Fig. 6 is 10mm for the acrylic wall, and this agrees with the value from Prantl's expression. However, the thickness of the boundary layer above the sound absorbing wall was about 28mm, and was about three times that in Prantl's expression because the sound-absorbing wall was made of fibered glass with a rough surface. It was clarified that the wall was necessary to obtain a wide measurement section and thus improve the uniformity of the flow.

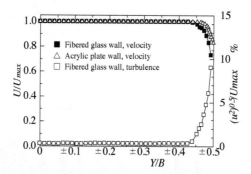

Figure 6. Flow velocity distribution and turbulence intensity in the test (measurement) section at a main flow velocity of U=28m/s

3.3. The relation between sound source and measurement position

It is known that the fluid-dynamic noise made from the circular cylinder placed into the air flow is a dipole sound. Since there is single directivity also in a microphone, it is important to understand the influence on the measurement result by the spatial relationship of a sound source and its microphone. Figure 7 shows the measurement result of the sound pressure level when varying the distance x between centers of a circular cylinder and a microphone in the range from 5mm to 95mm. Here, the airflow velocity in a test section was U=28m/s and the diameter of circular cylinder was 20mm. It is understood that the measured sound pressure level is almost the same. So, in measurement of acoustic frequency, distance between centers of circular cylinder and microphone was set to 50mm. Here, it is expected that the pressure fluctuation of a short-distance field is included in the sound pressure which will have been measured if the measurement position of sound is generally near from a sound source (circular cylinder). In this study, the distance between the circular cylinder and the microphone was narrower compared with the device arrangement for an ordinary sound measurement. Because the noise measurement of the flow in the fluid machine such as the gas turbines and jet engines is assumed, and the measurement of the fluid-dynamic sound caused by the flow around the object such as the supports and umbones installed in the pipeline and the duct is assumed, it becomes such arrangement. Therefore, the influence of the near field appears to be strong, making a quantitative evaluation of the sound pressure level more difficult. Resolving this is a clear challenge for future studies. The relationship between the position r_c by which the pressure fluctuation of a short-distance field can be disregarded now, and the minimum frequency f is given in $20\log(2\pi f r_c/a) >= 10$ dB (a is acoustic velocity) (Iida, 1996). Distance r_c between the circular cylinder and the microphone becomes 188mm-211mm because the range of center-to-center spacing x between the circular cylinder and the microphone in this experiment is 5mm-95mm. The obtained lower critical frequency f becomes 910Hz-812Hz. When the center-to-center spacing is assumed to be 50mm, distance r_c between two points becomes 194.5mm. The lower critical frequency f at that time is 880Hz.

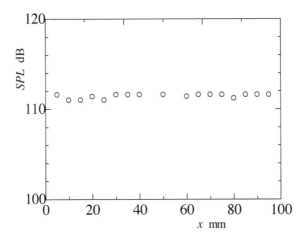

Figure 7. Measurement result for sound pressure level with a directivity check

3.4. Measurement result and verification of Acoustic frequency

The back ground noise with acoustic half-free space of a test section was measured by making airflow velocity in a test section into U=28m/s. The circular cylinder of various diameters
was installed in the test section, and frequency of a fluid-dynamic noise (acoustic frequency)
was measured. Figure 8 shows the results of the acoustic frequency analysis with back
ground noise (B.G.N.) in the measurement section and with a circular cylinder 20mm in diameter. The abscissa is frequency f, and the ordinate is the sound pressure level SPL in the
figure. When the circular cylinder is set up in the measurement part, a peak at one big
sound pressure level is obtained. At this time, the Strouhal number S (= $f d$ /U) calculated
from the frequency f (=275Hz) and air flow velocity U (=28m/s) is S=0.2. It is considered that
the microphone measures the acoustic frequency from the fluid oscillation based on Karman
vortex shedding. The frequency of the oscillating flow behind the circular cylinder was
measured using the hot wire anemometer for verification. Figure 9 shows the result of the
frequency analysis using the microphone and the hot wire anemometer. The abscissa is frequency f, and the ordinate is a sound pressure level made dimensionless by the maximum
value. In both measurement results, it is understood that one big peak is seen at the same
frequency. Therefore, it was established that the acoustic frequency measured by the microphone was a fluid oscillating frequency based on Karman vortex shedding from the circular
cylinder. This means that the fluid sound measured by making the acoustical free space can
be measured in an internal flow. And, this measurement technique is considered suitable for
the measurement of the fluid sound of an internal flow.

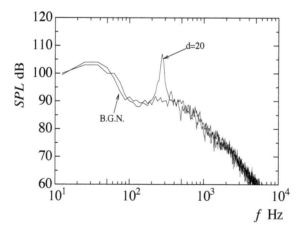

Figure 8. The frequency analysis of flow noise, U=28m/s, d=20mm

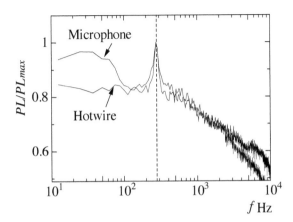

Figure 9. The relationship between the acoustic frequency and the fluid frequency due to vortex shedding

3.5. Comparison of measurement results with a blow-type wind tunnel

Figure 10 shows the variation of the peak frequency of the sound pressure level at the time of varying a circular cylinder diameter. The back ground noise is also shown for comparison. The abscissa is frequency f and the ordinate is a sound pressure level SPL. Increase of a cylinder diameter can see the tendency for a sound pressure level to increase and for peak frequency to decrease. The experimental result (Tomita et al., 1982) in the wind tunnel of a blow type with a half-opening type test section is shown in Fig. 11 for comparison with this experimental result. Although the variation of a sound pressure level or peak frequency to the variation of the diameter of the circular cylinder shows the same tendency, in each circular cylinder, one large peak and its harmonics component are seen, and spectrum distribution of the fluid-dynamic noise made when a circular cylinder is installed into an air current constitutes a larger sound pressure level than a back ground noise by the high frequency side which passed over the large peak. This suggests containing other sounds potential in not only the fluid-dynamic sound to be measured but also the flow noise. Therefore, it appears that the use of a blow-type wind tunnel with a half-open measurement section is rather inconvenient for measuring a sound effect. On the other hand, the results from a sealed-type measurement section of a suction-type wind tunnel becomes a sound pressure level that only the section of the frequency of the aimed fluid-dynamic sound is big as shown in Fig. 10, and the other frequency components are the same degree of the sound pressure level as the back ground noise. This is convenient for the examination of sound effects. The suction wind tunnel with a sealed-type measurement section can be expected to be a good measurement technique for examining sound effects.

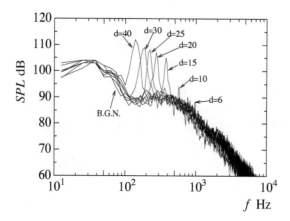

Figure 10. The characteristics of fluid-dynamic noise, in the case of present test section

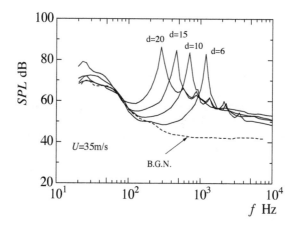

Figure 11. Characteristics of fluid-dynamic noise, in blow-type wind tunnel (Tomita et.al., 1982)

3.6. Effect of acoustic material and sound directivity

In order to verify the effect of the sound-absorbing material (fibered glass) in the measurement section, the acoustic frequency from the circular cylinder was measured. At this time, the microphone is set up from the bell mouse to 500mm upstream side by equal height to the circular cylinder installation position. The air flow velocity of the measurement section was U=28m/s. As a result, the measurement result differed according to the existence of the sound-absorbing material. Figure 12 shows the measurement results for circular cylinders 20mm, 25mm, 30mm, and 40mm in diameter when upper and lower sidewalls made of an acrylic board are used. Here, the back-ground noise (B.G.N.) is shown for comparison. The sound pressure levels for the peak frequency of each circular cylinder are small, and the peak frequency is twice the value of fluid oscillating frequency, based on the Karman vortex shedding shown in Fig. 10. On the other hand, the measurement result when sound-absorbing material is installed is shown in Fig. 13, relative to back-ground noise (B.G.N.) at circular cylinder diameters of 20mm-40mm. Two sound pressure peaks are seen in each figure. The first peak (1st peak) on the low frequency side is a Karman vortex shedding frequency, and the second peak (2nd peak) on the high frequency side is twice the Karman vortex shedding frequency. Moreover, the magnitude correlation of the two peaks is different in each circular cylinder. In the case of circular cylinders with diameters of 10mm, 15mm, and 30mm, the first peak (1st peak) on the low frequency side is larger. In the case circular cylinder diameters of 20mm, 25mm, and 40mm, the second peak (2nd peak) on the high frequency side is larger. It is shown that there is a change in the interference pattern of the sound wave in a vertical direction in the flow in the measurement section. The two peaks can also be observed in a blow-type wind tunnel with a half-open measurement section as shown in Fig. 11. In this case, however, the microphone is set up at and angle of 45 degrees and positioned 500mm behind the circular cylinder, aiming at the sound around the circular cylinder. The first peak (1st peak) on the low frequency side is always larger than the second peak (2nd peak) on the high frequency side in the magnitude correlation of the peak because of the position of the microphone and the directivity of the microphone. A comparison of the results between a suction-type and a blow-type wind tunnel with sound-absorbing material installed shows that the acoustical free space of an internal flow can become an acoustical free space similar to the case of an external flow. The sound is fluctuation of the pressure which transmits the inside of fluid, the size of the amplitude is the size of sound, and the height of oscillation frequency is the height of sound. The fluid force acts on the circular cylinder by the fluid fluctuation according to the vortex shedding from the circular cylinder. The oscillation of the fluid force can be divided into a lift component and a drag component, at a ratio of 1:2. The circular cylinder placed on the air flow is a source of two kinds of sound waves as two peaks are apparent in the fluid sound. When acrylic upper and lower sidewalls are used, the acoustical free space becomes the only flow direction. The sound by the oscillation in the direction of the lift is canceled by acoustical interference. Therefore, the peak frequencies of each circular cylinder shown in Fig. 12 are considered to be the oscillation a sound from the drag direction. On the other hand, when the sound-absorbing material is installed on the upper and lower sidewalls, the acoustical free space is two directional (a parallel direction and a vertical direction) for the flow. It is an acoustical free space similar to the blow-type wind tunnel with a half-open measurement section. Therefore, the sounds of the lift and drag oscillations are measured as shown in Fig. 13.

In general, because the oscillation amplitude of the lift is far larger than that of the drag, it is expected that the sound pressure level in the direction of the lift is far larger than the sound pressure of the drag direction. However, the sound from the oscillation of the drag direction is easily detected because the directivity microphone is located on the upstream side of the circular cylinder in this measurement, and the fluid-dynamic sound by the oscillation in the direction of the lift is not detected easily. In addition, because the interference pattern of the

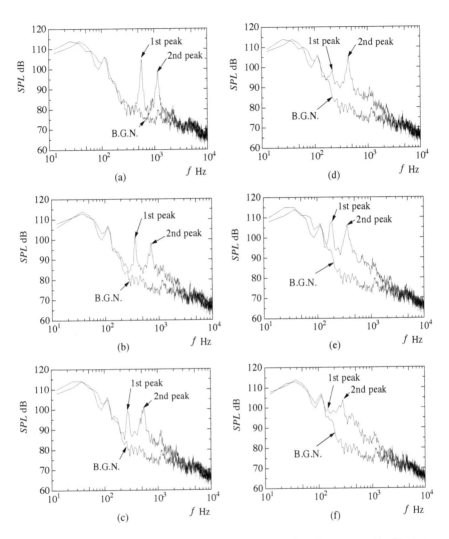

Figure 13. Characteristics of fluid-dynamic noise at the measurement point from the up-stream side of the test section, in the case of fibered glass wall; (a) cylinder diameter d is 10mm, (b) d=15mm, (c) d=20mm, (d) d=25mm, (e) d=30mm, (f) d=40mm

4. Conclusion

This study proposed a technique to measure the fluid-dynamic noise of an internal flow in a wind tunnel, and the fluid-dynamic noise from a circular cylinder placed on the air flow of a suction-type wind tunnel with a sealed-type measurement section with sound-absorbing material (fibered grass) was measured. The following conclusions were obtained.

1. The acoustic performance and fluid-dynamic performance of a test wind tunnel were good. The following results were obtained for the performance of the test wind tunnel. The noise in the blower room is effectively intercepted. The position of the sound source and the microphone are not influenced by directivity. The uniformity of the flow of the measurement section narrows when sound-absorbing material is used for the measurement section of the test wind tunnel.

2. The following results were obtained from installing sound-absorbing material in the measurement section. The acoustical free space can be made from the closed space. When the surface of the microphone was arranged and set up on the surface of the sound-absorbing material, the measurement of the fluid sound of an internal flow became possible without any disarrangement of the flow-field.

3. The acoustic frequency measured by the microphone was confirmed to have a frequency based on the fluid oscillation caused by the Karman vortex shedding measured with the hot-wire anemometer.

4. The following results were obtained when a comparison was made with the results from a blow-type wind tunnel. The aimed acoustic frequency was measured by the large sound pressure level. Other frequency elements were the same degrees of the sound pressure level as the back ground noise. It has been understood that such a result was convenient when a sound effect was examined.

5. When an acoustical effect was examined, it was understood that the following consideration is necessary. The distance between the sound source and the microphone must be set in consideration of the influence of the pressure fluctuation of the near-field. The lower bound frequency must be understood. The microphone must be arranged in consideration of the sound directivity with the sound source.

6. From the results outlined in (2)-(4), this present measurement technique is considered to be a technique useful for the measurement of the fluid sound of an internal flow.

Nomenclature

a: acoustic velocity, m/s

B: height of test section, m

B.G.N.: back ground noise, dB

d: diameter of circular cylinder, m

f: frequency, Hz

I: intensity of sound, W/m^2

I_0: intensity of sound of standard, W/m^2

L: level of sound intensity, dB

L_{IN}: level of sound intensity inside fan room, dB

L_{OUT}: level of sound intensity outside fan room, dB

PL/PL_{max}: dimensionless sound pressure level, non-dimensional

r_c: critical distance, m

S: Strouhal number, non-dimensional

SPL: sound pressure level, dB

U: main flow velocity, m/s

U_{max}: maximum flow velocity, m/s

u: component of fluctuating flow velocity, m/s

x: measurement position in test section, m

x: coordinate component (flow direction), m

Y: measurement position in test section, m

y: coordinate component (vertical direction), m

z: coordinate component (horizontal direction), m

δ: thickness of boundary layer, m

ν: kinematic viscosity of air, m^2/s

Author details

Yoshifumi Yokoi

National Defense Academy of Japan, Japan

References

[1] Fujita, H. (1994). The Present Condition and View for the Basic-study About the Flow and Sound Controls. *Proceedings of the 72nd the JSME Fall Meeting of the Japan Society of Mechanical Engineers*, pp. 360-363, Tokyo, Japan, October, 1994

[2] Fujita, H. (1996). Experimental Study on Aerodynamic Noise Generated from two-Dimensional Models (1st Report, Study on Wind Tunnel Wall Effect and Wall Mate-

rials), *Transactions of the Japan Society of Mechanical Engineers, Series B*, Vol. 62, No. 593, (January 1996), pp. 187-193

[3] Iida, A., Fujita, H., Kato, C., & Otaguro, Y. (1996). Experimental Investigation of the Generation Mechanism of Aerodynamic Noise (2nd Report, On Correlation Between Surface Pressure Fluctuation and Aerodynamic Sound Radiated from a Circular Cylinder), *Transactions of the Japan Society of Mechanical Engineers, Series B*, Vol. 62, No. 604, (December 1996), pp. 4160-4167

[4] Mochizuki, O., & Maruta, Y. (1996). *Introduction of Fluid Sound Engineering*, Asakura-shoten, ISBN 4-254-23088-5, Tokyo, Japan

[5] Tomita, S., Suzuki, S., Inagaki, S., Yokoyama, T., Kobayashi S., & Tsukamoto, G. (1982). The Study about Fluid Sound from a Cylinder in a Uniform Flow (1st Report, in the Case of Circular Cylinders), *Proceedings of the Thohoku Branchi Meeting of the Japan Society of Mechanical Engineers*, Senday, Japan, October, 1984

Statistical Analysis of Wind Tunnel and Atmospheric Boundary Layer Turbulent Flows

Adrián Roberto Wittwer,
Guilherme Sausen Welter and Acir M. Loredo-Souza

Additional information is available at the end of the chapter

1. Introduction

Many studies on wind engineering require the use of different types of statistical analysis associated to the phenomenology of boundary layer flows. Reduced Scale Models (RSM) obtained in laboratory, for example, attempt to reproduce real atmosphere phenomena like wind loads on buildings and bridges and the transportation of gases and airborne particulates by the mean flow and turbulent mixing. Therefore, the quality of the RSM depends on the proper selection of statistical parameters and in the similarity between the laboratory generated flow and the atmospheric flow.

The turbulence spectrum is the main physical parameter used to compare the velocity fluctuation characteristics of atmospheric and laboratory flows in Wind Load Modeling (WLM). This is accomplished by fitting experimental spectra to some functional form, *e.g.*, von Kármán, Harris or Batchelor-Kaimal formula, and then creating dimensionless turbulence spectra in accordance with a similarity theory [1, 2, 3]. The objective behind the use of a similarity theory is that the dimensionless spectra of atmospheric and laboratory flows collapse, if the dimensionless spectra were constructed by appropriate parameters [4].

This classical spectral comparison is commonly used in WLM [5]. However, some difficulties, related to the determination of the inertial range extent, choice of characteristic velocity and length scale parameters and possible effects due to the finiteness of the Reynolds number arise in wind tunnel studies, specially, when simulations are performed at low velocities [6].

Considering this scenario, a complementary study taking into account the use of local scale based Reynolds number, inertial and dissipation range characteristic scales, control of sampling frequency and post-processing filtering is proposed. Selected data sets obtained under distinct configurations of three wind tunnels, a smooth pipe and atmospheric boundary layer are used. In addition, a different class of spectral representation proposed by Gagne et al. [7], which is based on local similarities and compatible with the multifractal formalism, is compared to traditional approaches.

2. Atmospheric boundary layer flows and wind tunnel flow simulation

The atmospheric boundary layer is the lowest part of atmosphere. Effects of the surface roughness, temperature and others properties are transmitted by turbulent movement in this layer. Under conditions of weak winds and very stable stratification, turbulent exchanges are very weak and the atmospheric boundary layer is called surface inversion layer [8]. A distinction is usually made between atmospheric boundary layer over homogeneous and non-homogeneous terrain. In this last situation, the boundary layer is not well defined, and topographical features could cause highly complex flows.

The depth of the atmospheric boundary layer varies with the atmospheric condition, but it is typically 100 m during the night-time stable conditions and 1 km in daytime unstable or convective conditions. A detailed description related to the wind characteristics associated to the neutral condition is made by Blessmann [3] in his book on Wind in Structural Engineering. Several similarity theories have been proposed for different atmospheric stability conditions. Near the surface, from dimensional arguments the analysis leads to the Prandtl logarithmic law, Eq. (1), in the case of a neutral boundary layer:

$$\frac{U(z)}{u^*} = \frac{1}{0.4} \ln \frac{z - z_d}{z_0} \qquad (1)$$

Where U is the mean velocity, u^* is the friction velocity, z_0 is know as the roughness height and z_d is defined as the zero-plane displacement for very rough surface. The depth of a wind tunnel boundary layer is defined as the height where mean velocity reaches 0.99 of the free stream velocity. This definition is used to characterize atmospheric flow simulations.

Wind tunnels are designed to obtain different air flows, so that similarity studies can be performed, with the confidence that actual operational conditions will be reproduced. Once a wind tunnel is built, the first step is the evaluation of the flow characteristics and of the possibility of reproducing the flow characteristics for which the tunnel was designed. Many evaluation studies of wind tunnels are presented in the open literature. Some of which are the work of Cook [9] on the wind tunnel in Garston, Watford, UK, the presentation of the closed-return wind tunnel in London [10] and the Oxford wind tunnel, UK [11], the characterization of the boundary layer wind tunnel of the UFRGS, Brazil [12] and of the Danish Maritime Institute, Denmark [13].

Wind tunnel modeling of atmospheric boundary layer is generally oriented to neutrally stable flows. Modeling of stratified boundary layer is more difficult to implement and less used in wind tunnel tests. Similarity criteria imply that a set of non-dimensional parameters should be the same in model and prototype. In general, the flow is governed by the boundary conditions and the Rossby, Reynolds, Strouhal, Froude, Eckert and Prandtl numbers, but in most of the situations of practical importance the effects of several non-dimensional numbers can be neglected. Later studies in atmospheric boundary layer simulations attempted to reproduce as closely as possible the mean velocity distribution and turbulence scales of the atmospheric flow. This is made by non-dimensional comparisons of mean and fluctuating velocity measurements in the wind tunnel flow and atmospheric data.

In general, wind tunnel evaluation is performed at the highest flow velocity, the results being presented in terms of mean velocity distributions, turbulence intensities and scales. However, many simulations are performed at low velocities to evaluate some specific problems. This is the case of laboratory simulation of dispersion problems [14] and transmission line modeling [15].

Boundary-layer simulations are performed with help of grids, vortex generators and roughness elements, to facilitate the growth of the boundary layer and to define the mean velocity profile. This is used in the most applied simulation methods, namely the full-depth simulation [16] and part-depth simulation [17]. The use of jets and grids is also applied [12].

The "Jacek Gorecki" wind tunnel, located at the Universidad Nacional del Nordeste, UNNE at Resistencia (Chaco), Argentina, is a low velocity atmospheric boundary-layer wind tunnel, built with the aim to perform aerodynamic studies of structural models. The atmospheric boundary layer is reproduced with help of surface roughness elements and vortex generators, so that natural wind simulations are performed. Fig. 1 shows a view of the "Jacek Gorecki" wind tunnel, which is a 39.56 m long channel. The air enters through a contraction, passing a honeycomb prior to reach the test section, which is a 22.8 m long rectangular channel (2.40 m width, 1.80 m height). Two rotating tables are located in the test section to place structural models. Conditions of zero pressure gradient boundary layers can be obtained by vertical displacement of the upper wall. The test section is connected to the velocity regulator and to the blower, which has a 2.25 m diameter and is driven by a 92 kW electric motor at 720 rpm. A diffuser decelerates the air before leaving the wind tunnel.

In this wind tunnel, many models of atmospheric boundary layer were implemented. In general, the simulation of natural wind on the atmospheric boundary layer was performed by means of the Counihan and Standen methods [18, 19, 20]. To illustrate this type of flow model, an example of full-depth Counihan simulation with velocity distributions corresponding to a class III terrain is presented. According to Argentine Standards CIRSOC 102 [21], this type of terrain is designed as "ground covered by several closely spaced obstacles in forest, industrial or urban zone". The mean height of the obstacles is considered to be about 10 m, while the boundary layer thickness is $z_g = 420$ m. The power law for velocity distribution is given by

$$\frac{U(z)}{U(z_g)} = \left(\frac{z}{z_g}\right)^{\alpha} \tag{2}$$

with suitable values for the exponent a between 0.23 and 0.28 [3]. This law is of good application in neutral stability conditions of strong winds, typical for structural analysis. For this
Counihan full-depth simulation, where the complete boundary-layer thickness is simulated,
four 1.42 m (Hv) high elliptic vortex generators and a 0.23 m (b) barrier were used, together
with prismatic roughness elements placed on the test section floor along 17 m (l) (see Fig. 2).
The wind tunnel test section and the simulation hardware are shown in Fig. 3.

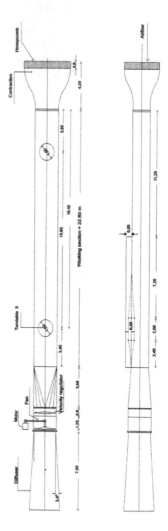

Figure 1. "Jacek Gorecki" Wind Tunnel at UNNE.

In this work, measurements of wind velocity realized in three different wind tunnels will be used for the spectral analysis. The *"Jacek Gorecki"* wind tunnel [19] described above, the *"TV2"* wind tunnel of the Laboratorio de Aerodinámica, UNNE, smaller, also an open circuit tunnel, and the closed return wind tunnel *"J. Blessmann"* of the Laboratório de Aerodinâmica das Construçoes, Universidade Federal de Rio Grande do Sul, UFRGS [12].

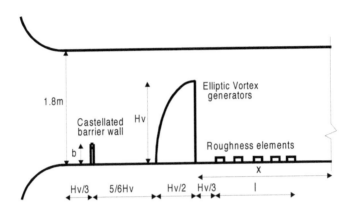

Figure 2. Arrangement of Counihan simulation hardware.

Figure 3. Test section and full-depth Counihan vortex generators.

3. Mean flow evaluation of an atmospheric boundary layer simulation

The above described Counihan simulation is used to illustrate the mean flow evaluation. In this case, mean velocity measurements were performed by means of a pitot-Prandtl tube connected to a Betz manometer. Velocity and longitudinal velocity fluctuations were measured by a constant temperature hot wire anemometer, with a true-RMS voltmeter, using low and high-pass analogical filters. Data acquisition of hot wire signals was made by means of an A/D board connected to a personal computer. Uncertainty associated with the measured data depends of the hot wire resolution and the calibration system. In this case, an uncertainty order of ± 3 % was determined at high velocity measurements.

Previously to simulate ABL flows, an empty tunnel flow evaluation was realized. Mean velocity profiles were measured along a vertical line on the center of the rotating table 2. The boundary layer has a thickness of about 0.3 m and the velocity values have a maximal deviation of 3 %, by taking the velocity at the center of the channel as reference. Turbulence intensity distribution at the same locations shows values around 1% outside the boundary layer increasing, as expected, inside the boundary layer. Reference velocity at the center of the channel for empty tunnel tests was 27 m/s and the resulting Reynolds number $3.67×10^6$.

Once the empty tunnel evaluation was over, the mean flow of the full-depth boundary layer simulation was analyzed. Measurement of the mean velocity distribution was made along a vertical line on the center of rotating table 2 and along lines 0.30 m to the right and left of this line. Fig. 4 shows the velocity distribution along the central line. Flow characteristics are presented in Table 1. There is a good similarity among the velocity profiles given by the values of the exponent α obtained. Turbulence intensity distribution at the same locations is shown in Fig. 5. The values are lower than those obtained by Cook [4] and by using Harris-Davenport formula for atmospheric boundary layer [3]. Values are reduced as the distance from the lower wall is increased.

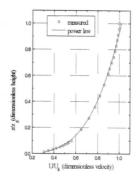

Figure 4. Vertical mean velocity profile – experimental values and power law fit.

These mean velocity and turbulence intensity vertical profiles show a typical evaluation of the boundary layer mean flow applied to wind load studies. Similar analysis was made for other authors to different wind tunnel simulations. Some works include vertical profiles of longitudinal turbulence scales [12, 20]. When dispersion problems are analyzed and physic atmospheric research studies in wind tunnel are development the mean flow evaluation is usually realized utilizing the logarithmic expression, Eq. (1). A simple method to fit experimental values of mean velocity to the logarithmic law is presented by Liu et al. [22]. The characteristic parameters u^* and z_0, friction velocity and roughness height, respectively, are used to evaluate critical Reynolds number values on low velocity tests for wind tunnel dispersion studies [23].

	$y = 0$	$y = 0.30$ m	$y = -0.30$ m
z_g[m]	1.164	1.164	1.164
U_g[m/s]	27.51	28.18	27.76
a	0.270	0.265	0.270

Table 1. Flow characteristics for full-depth boundary layer simulation.

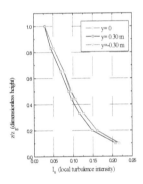

Figure 5. Vertical turbulence intensities profiles.

4. Energy spectra and structure functions in boundary layer flows

Atmospheric data come from anemometers frequently located 10 m height. These values contain climatic system contributions and components of the boundary layer itself. That is, measured data include wind velocity variations corresponding to time scales from some hours to fractions of one second. Usually power spectra are employed to analyze these atmospheric records. The Van der Hoven spectrum, obtained in Brookhaven, Long Island, NY, USA [24], represents the energy of the longitudinal velocity fluctuation on the complete frequency domain. Two peaks can be distinguished in this spectrum, one corresponding to the 4-day period or 0.01 cycles/hour (macro-meteorological peak), and another peak between the periods of 10 minutes and 3 seconds associated to the boundary layer turbulence (micro-meteorological peak). A spectral valley, with fluctuations of low energy, is observed between the macro and micro-meteorological peak. This region is centered on the period of 30 minutes and allows dividing the mean flow and the velocity fluctuations. This spectral characteristic confirms that interaction between climate and boundary layer turbulence is negligible and permits considering both aspects independently.

Velocity fluctuations with periods lower than one hour define the micro-meteorological spectral region or the atmospheric turbulence spectrum. Interest of wind load and dispersion problems is concentrated on this spectral turbulence region. In 1948 von Kármán suggested an expression for the turbulence spectrum with which his name is related, and 20 years later this spectral formula started to be used for wind engineering applications. Some deficiencies in fitting data measured in atmospheric boundary layer were pointed later and Harris [5] shown a modified formulae for the von Kármán spectrum.

According ESDU [3], the von Kármán formula for the dimensionless spectrum of the longitudinal component of atmospheric turbulence is:

$$\frac{fS_u}{\sigma_u^2} = \frac{4X_u(z)}{\left[1 + 70{,}78X_u(z)^2\right]^{5/6}} \tag{3}$$

where S_u is the spectral density function of the longitudinal component, f is the frequency in Hertz and σ_u^2 is the variance of the longitudinal velocity fluctuations. The dimensionless frequency $X_u(z)$ is $fL(z)/U(z)$, being L the integral scale. This spectrum formula satisfies the Wiener-Khintchine relations between power spectra and auto-correlations and provides a Kolmogorov equilibrium range in the spectrum. However, the von Kármán expression provides no possibility to fit other measured spectral characteristics [5].

Two situations of spectral analysis of boundary layer flow are presented next from different wind tunnel studies and atmospheric data. These cases resume a typical spectral evaluation of a boundary layer simulation and a spectral comparison of different boundary layer flows. Finally, a discussion of the use of structure functions applied to the analysis of velocity fluctuations is presented.

4.1. Spectral evaluation of a wind tunnel boundary layer simulation

A first example of spectral analysis is that corresponding to the Counihan boundary layer simulation described on previous section. Longitudinal velocity fluctuations were measured by the hot wire anemometer system and the uncertainty associated with the measured data is the same as previously mentioned. In this case, spectral results from longitudinal velocity fluctuations were obtained by juxtaposing three different spectra from three different sampling series, obtained in the same location, each with a sampling frequency, as given in Table 1, as low, mean and high frequencies. The series were divided in blocks to which an FFT algorithm was applied [25]. In Fig. 6, four spectra obtained at height z=0.233, 0.384, 0.582 and 0.966 m are shown. Values of the spectral function decrease as the distance from the tunnel floor z is increased. An important characteristic of the spectra is the presence of a clear region with a -5/3 slope, characterizing Kolmogorov's inertial sub-range.

The comparison of the results obtained through the simulations with the atmospheric boundary layer is made by means of dimensionless variables of the auto-spectral density fS_u/σ_u^2 and of the frequency $X_u(z)$ using the von Kármán spectrum, given by the expression of Eq. (3). Kolmogorov's spectrum will have, therefore, a -2/3 exponent instead of -5/3. The comparison was realized for spectra measured at different heights, but only is presented the spectrum obtained at z = 0.233 m (Fig. 7). The agreement is very good, except for the highest frequencies affected by the action of the low-pass filter.

	Low frequency	Mean frequency	High frequency
Sampling frequency [Hz]	300	900	3000
Low-pass filter [Hz]	100	300	1000
High-pass filter [Hz]	0.3	0.3	0.3
Sampling time [s]	106.7	35.6	10.7
Bandwidth [Hz]	1.132	3.516	11.719

Table 2. Data acquisition conditions for spectral analysis.

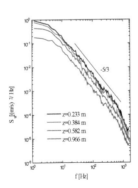

Figure 6. Power spectra of the longitudinal velocity fluctuation for a boundary layer simulation.

This evaluation was realized at high velocity ($U_g \approx 27$ m/s) being the resulting Reynolds number value of Re $\approx 4 \times 10^6$. The juxtaposing technique used to improve the spectral resolution is today unnecessary because of the fact that is possible to utilize a large sample size. However, sample series were limited to 32000 values for this analysis and three spectra were juxtaposed.

A scale factor of 250 for this boundary layer simulation was obtained through the procedure proposed by Cook [4], by means of the roughness length z_0 and the integral scale L_u as parameters. The values of the roughness length are obtained by fitting experimental values of velocity to the logarithmic law of the wall, while integral scale is given by fitting the values of the measured spectrum to the design spectrum.

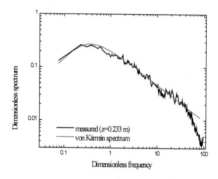

Figure 7. Comparison of the dimensionless spectrum obtained at z = 0.233 m and the von Kármán spectrum.

4.2. Spectral comparison of different boundary layer flows

A second study based on results of different boundary layer flows was realized. Measurements of the longitudinal fluctuating velocity obtained in three different wind tunnels were selected for this analysis. All selected velocity samples correspond to neutral boundary layer flow simulations developed in appropriate wind tunnels. The analysis was complemented using measurements realized in a smooth tube flow and in the atmosphere.

Wind tunnel and smooth tube measurements were realized by a constant hot-wire anemometer previously described. Atmospheric data were obtained using a Campbell 3D sonic anemometer [26], for which the resolution is 0.01 m/s for velocity measurements. Table 3 indicates a list of sampling characteristics, being z the vertical position (height), U the mean velocity, σ_u^2 the variance of fluctuations velocity, f_{acq} the acquisition frequency, L_u the integral scale and Re_L the Reynolds number associated to L_u.

One of the three wind tunnels used to obtain the wind data employed in this experimental analysis is the "*Jacek Gorecki*" wind tunnel described on a previous section. The second is the

"*TV2*" wind tunnel of the Laboratorio de Aerodinámica, UNNE, too. The "*TV2*", smaller, is also an open circuit tunnel with dimensions of 4.45×0.48×0.48m (length, height, width). The study was complemented by the analysis of measurements realized on atmospheric boundary layer simulations performed in the closed return wind tunnel "Joaquim Blessmann" of the Laboratório de Aerodinâmica das Construçoes, Universidade Federal de Rio Grande do Sul, UFRGS [12]. The simulations of natural wind on the atmospheric boundary layer were performed by means of the Counihan [16] and Standen [17] methods, with velocity distributions corresponding to a forest, industrial or urban terrain. The tube measurement was obtained in the centre of a 60 mm diameter smooth tube. Atmospheric data were obtained in a micrometeorological station located at Paraiso do Sul, RS, Brasil [26, 27].

	z [m]	U [m/s]	σ_u^2 [m²/s²]	f_{acq}[Hz]	L_u [m]	Re_L
Smooth tube	0.03	38.89	1.63	16	0.034	8.83×10⁴
Atmosphere	10.00	4.51	3.32	16	36.30	1.09×10⁷
Blessman WT-LV	0.15	3.18	0.19	1024	0.51	1.08×10⁵
Gorecki WT-LV	0.21	2.97	0.26	1024	0.26	5.16×10⁴
Gorecki WT-HV[+]	0.21	16.77	7.55	2048	0.51	5.71×10⁵
TV2 WT-LV	0.04	0.68	0.03	900	0.07	3.18×10³
TV2 WT-HV	0.04	11.69	4.92	3000	0.11	8.59×10⁴

Table 3. Measurement characteristics for spectral analysis.

The measurements realized in the J. Gorecki wind tunnel at high velocity were used to analyze the sampling effects on the spectral characteristics. Five different samplings were realized for measurements Gorecki WT-HV[+] at z= 0.21 m. Sampling characteristics like frequency acquisition f_{acq}, low pass frequency f_{lp} and sampling time t_s are indicated in Table 4. Resulting superposed spectra are shown in Fig. 8 where it is possible to see a good definition of the inertial sub-range (-5/3 slope) and the effect of the low pass filter.

Samples	Sp1	Sp2	Sp3	Sp4	Sp5
f_{acq}[Hz]	4096	2048	1024	8192	16348
f_{lp}[Hz]	3000	1000	300	3000	10000
t_s[s]	30	60	120	15	7.5

Table 4. Measurement characteristics for analysis of sampling effects.

Fig. 9 shown spectral density functions S_u corresponding to measurements indicated in Table 3. High frequencies in the atmosphere spectrum correspond to low frequencies in the smooth pipe. The same spectra in dimensionless form are presented in Figs. 10 and 11. The frequency is non dimensionalised by fL_u/U in Fig. 10 and by fz/U in Fig. 11, according to parameters usually employed in wind engineering. In general, preliminary results permit verifying the good behavior of the wind tunnel spectra and a good definition of the inertial range (slope -5/3). The inertial sub-region is narrower for low velocity measurements (LV).

Spectral special features in smooth tube and atmosphere appear in Fig. 9 and in the dimensionless comparison too (Figs. 10 and 11). This particular behavior is a product of the uniform flow in the centre of the smooth tube, that is, not a boundary layer flow is being analyzed. In the atmospheric flow case, this type of behavior is possibly due to the existence of a convective turbulence component at low frequencies because of that atmospheric stability is not totally neutral. This behavior was verified in the case of measurements realized in near-neutral atmosphere. The existence of a low frequency convective component was detected in three dimensional measurements obtained at the atmosphere [28]. The aliasing effect is perceived at high frequencies due to high pass filter is not used for sample acquisition of atmospheric data.

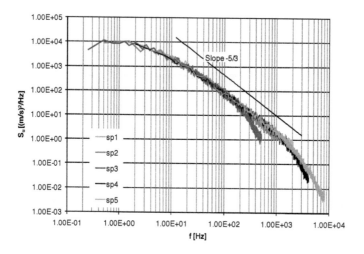

Figure 8. Spectral superposition for different sampling.

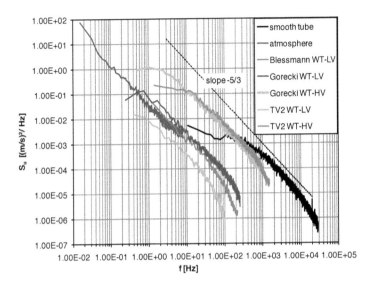

Figure 9. Power spectra for measurements indicated on Table 3.

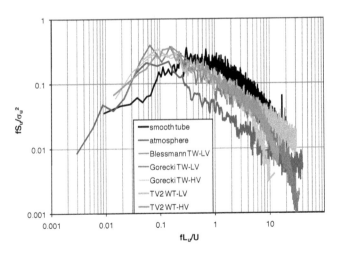

Figure 10. Comparison of dimensionless spectra using fL_u/U.

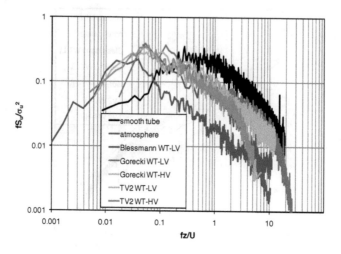

Figure 11. Comparison of dimensionless spectra using fz/U.

The superposition technique allows defining precisely the sub-inertial range and extending the frequency analysis interval. Besides, it is possible defining adequately the sampling characteristics and optimizing the measuring time. In general, the spectral comparison realized using fL_u/U (Fig. 10) indicates better coincidence [27, 28]. However, the analysis realized up to now is preliminary and it should be studied in depth. For example, the methods for the parameter L_u calculation should be analyzed, the application of other parameters to obtain the dimensionless frequency at smaller scales and other measurements must be analyzed looking for the improvement of the scale modeling.

A different approach to analyze velocity fluctuations will be presented below. This is based on the high order moments of velocity increments. Small scales to characterize the boundary layer flows will be used and a new representation of energy spectra will be evaluated.

4.3. Statistical moments of velocity fluctuations

Previous type of spectral analysis is usually employed in Wind engineering. The following study is realized using velocity structure functions of turbulent boundary layer flow. These statistical moments are utilized by atmospheric physical researchers. The approach considers scales smaller than the integral scale L_u and, therefore is presumably more suitable for applications to turbulent diffusion studies. Apart from integral scales, the mean dissipation rate, the Kolmogorov and Taylor micro-scales could be obtained. On other hand, results from this type of study can be employed to analyze the Kolmogorov constant and, indirectly, for application to pollution dispersion models [30, 31].

Kolmogorov's laws for locally isotropic turbulence [32, 33] were originally derived for structure functions from the von Kármán-Howarth-Kolmogorov equation [34],

$$S_3(r) = -\frac{4}{5}\varepsilon r + 6\nu \frac{d}{dr} S_2(r) \tag{4}$$

valid for $r \ll L_u$ in the limit of very large Reynolds number, where $S_p(r) = \langle [u(x+r) - u(x)]^p \rangle$ is the structure function of order p, ν is the kinematic viscosity, ε is the mean dissipation rate, and $\langle \cdot \rangle$ represents statistical expectation operator.

Kolmogorov deduced the following relations for second and third-order structure functions:

$$S_2(r) = C(\varepsilon r)^{2/3} \tag{5}$$

$$S_3(r) = -\frac{4}{5}\varepsilon r, \tag{6}$$

valid for $\eta \ll r \ll L_u$, where $\eta = (\nu^3/\varepsilon)^{1/4}$ is the Kolmogorov microscale, and $C \approx 2$ is the Kolmogorov constant [29, 34].

The third-order structure function Eq.(6), also known as the four-fifths law, is straightforwardly obtained from Eq.(4) since, for very large Reynolds number, the second term in the right hand side of Eq.(4) can be neglected. The four-fifths law is of special interest in the statistical theory of turbulence because, besides being an exact relation, it allows a direct identification of the mean dissipation energy per unit mass with the mean energy transfer across scales [35].

The two-thirds law Eq. (3), on the other hand, is not an exact relation; it was obtained using dimensional arguments and introducing a nondimensional constant that should be empirically determined. The second-order structure function provides information about the energy content in all scales smaller than r. Moreover, the famous Kolmogorov energy spectrum $E(k) = C_k \varepsilon^{2/3} k^{-5/3}$ is derived from Eq. (5).

Table 5 shows the results of the analysis for four experiments selected from the analysis described in section 4.2. The distinct columns report the mean wind speed U, height z, inertial range (r_a, r_b), integral scale L_u, mean dissipation rate ε, Kolmogorov microscale η, Taylor's microscale based Reynolds number Re_λ. The mean dissipation rate, ε, was determined by the best fit of $S_3(r)$, Eq. (2), in the inertial range. The Kolmogorov microscale was computed by $\eta = (\nu^3/\varepsilon)^{1/4}$ and Taylor's microscale based Reynolds number $Re_\lambda = \sigma_u \lambda/\nu$ was computed from $\lambda = [\sigma_u^2/\langle(\partial_x u)^2\rangle]^{1/2}$, where $\langle(\partial_x u)^2\rangle$ was indirectly estimated with the aid of the isotropic relation $\varepsilon = 15\nu\langle(\partial_x u)^2\rangle$.

	z [m]	U [m/s]	(r_a, r_b)[cm]	L_u [m]	ε [m²/s³]	η [mm]	Re_λ
Smooth tube	0.03	38.89	0.35-1.10	0.034	52.9	0.08	174
Atmosphere	10.00	4.51	30-600	36.30	0.045	0.51	13141
Gorecki WT-HV	0.21	16.77	2.0-9.0	0.39	33.0	0.10	1311
TV2 WT-HV	0.04	11.69	0.3-2.0	0.13	48.8	0.09	629

Table 5. Main turbulence characteristics from laboratory and atmospheric turbulence data.

Experimental evaluations of second and third-order structure functions for the J. Gorecki wind tunnel are shown in Fig. 12. In the K41 picture, the estimation of the second-order structure function constant is reduced to an estimation of the skewness $S = S_3(r)/(S_2(r))^{3/2}$; however, differently from $S_2(r)$, $S_3(r)$ displays some noise (Fig. 12). This behavior is observed in all datasets.

One immediate consequence of the similarity arguments assumed in K41 is that graphical representation of distinct turbulence spectra should collapse in a single-curve after a proper normalization with characteristic velocity and length scales. Another consequence, which follows from dimensional analysis, is the scaling $S_p(r) \approx r^{p/3}$ for a structure function of order p, with $\eta \ll r \ll L_u$. However, inertial range physics has been proved to be much more complex than previously assumed in the K41. A remarkable consequence of this complexity, which has close relation with the small scale intermittency phenomenon [33], is the existence of anomalous scaling concerning structure functions exponents, $S_p(r) \approx r^{\zeta_p}$, where ζ_p is non linear function of p. The multifractal formalism was then introduced by Parisi and Frisch in order to provide a robust framework, allowing the analysis and interpretation for a general class of complex phenomena presenting anomalous scaling.

One important difference between the multifractal interpretation of turbulence and the (monofractal) K41 theory is the assumption of a local similarity scaling for small scales. The global scaling similarity assumed in the K41 theory is still at the core of the most wind tunnel and atmospheric turbulence modeling [5, 28, 34]. The local scaling similarity ideas of the multifractal formalism, on the other side, provide a new vocabulary, enabling interpretation and comparison of diverse multiscale phenomena. Although the multifractal formalism has been used in many areas of applied physics, does not share the same popularity in the fields of engineering.

According to the multifractal universality [36], a single-curve collapse of distinct experimental turbulence spectra is obtained by plotting $\log E(k)/\log Re$ against $\log k/\log Re$, after having properly normalized $E(k)$ and k. On the other hand, an alternative similarity plot has been proposed by Gagne et al. [7] based on an intermittency model, but still compatible with the multifractal formalism. These authors propose that a better merging of experimental spectra can be obtaining by plotting $\beta \log\left(aE(k)(\varepsilon v^5)^{-1/4}\right)$ against $\beta \log\left(ck(v^3/\varepsilon)^{1/4}\right)$, with $\beta = 1/\log(Re_\lambda/R^*)$, where Re_λ is the Taylor scale based Reynolds number and the empirical constants $R^* = 75$, $a = 0.154$, and $c = 5.42$ were determined to provide the best possible superposition in their dataset.

In Fig. 13 the plot proposed by Gagne et al. [7] is presented for laboratory and atmospheric turbulence data. Despite the fact that data comprise very different scales, $L_u \approx 10^2$ m for atmospheric data, and $L_u \approx 10^{-1}$ m for smooth pipe, the merging of spectra is reasonably good, also regarding the fact that the originally proposed empirical constants have been used in the present dataset.

In this representation the slopes remain unchanged, but the extent of inertial range presumably has the same length for all spectra. Although a solid ground for the physics behind the representation is lacking, it is clear that the properties provided by such a representation can be very useful for physical analysis and modeling of turbulence.

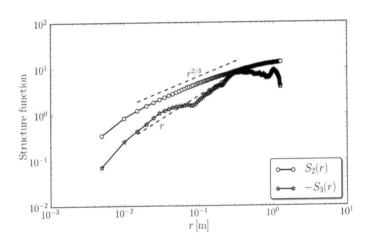

Figure 12. Second and third-order structure functions for the Gorecki WT measurement.

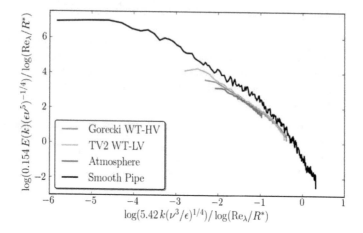

Figure 13. Single-curve spectral collapse from laboratory and atmospheric turbulence data (Table 5), as proposed by Gagne et al. [7].

5. Concluding remarks

Fully developed turbulence measurements from the laboratory and the atmospheric boundary layer encompassing a wide range of Reynolds number were analyzed in this study. First, a typical spectral evaluation of a boundary layer simulation was presented. The spectral agreement is very good and the wind simulation can be considered adequate for wind load modeling.

Next, a spectral dimensionless comparison of different boundary layer flows was realized by usual parameters in wind engineering. Measurements of the longitudinal fluctuating velocity obtained in different wind tunnels, a smooth tube and the atmosphere were selected. An analysis of sampling effects was realized and some limitations on this classical spectral comparison were established.

Finally, a discussion of the use of structure functions to investigate turbulent boundary layer flows was proposed. Turbulent scales smaller than the integral scale were determined and the behavior of second and third-order structure functions were analyzed. A single-curve collapse of distinct experimental spectra was obtained. This type of analysis should be applied to verify boundary layer flows at low speed used for dispersion modeling. Time scales for fluctuating process modeling could be improved too by applying this analysis method.

Acknowledgements

The authors acknowledge the atmospheric physics group from Universidade Federal de Santa Maria for sharing their atmospheric boundary layer measurements. One of us (GSW) is supported by a PCI scholarship provided by the Brazilian research agency CNPq.

Author details

Adrián Roberto Wittwer[1], Guilherme Sausen Welter[2] and Acir M. Loredo-Souza[3]

*Address all correspondence to: a_wittwer@yahoo.es

1 Facultad de Ingeniería, Universidad Nacional del Nordeste, Argentina

2 Laboratório Nacional de Computação Científica, Brasil

3 Universidade Federal de Rio Grande do Sul, Brasil

References

[1] Kaimal, J. C., Wyngaard, J. C., Izumi, Y., Cote, O. R., "Spectral characteristics of sur-face-layer turbulence", Quart. J. R. Met. Soc. 1972, 98: 563-589.

[2] Kaimal, J. C., "Atmospheric boundary layer flows: their structure and measurement", Oxford University Press, Inc., New York, 1994.

[3] Blessmann, J., "O Vento na Engenharia Estrutural", Editora da Universidade, UFGRS, Porto Alegre, Brasil, 1995.

[4] Cook, N. J., Determination of the Model Scale Factor in Wind-Tunnel Simulations of the Adiabatic Atmospheric Boundary Layer, Journal of Industrial Aerodynamics, 1978, 2: 311-321.

[5] Harris, R. I., "Some further thoughts on the spectrum of gustiness in strong winds", J. of Wind Eng. & Ind. Aerodyn. 1990, 33: 461-477.

[6] Isymov, N., Tanaka, H., "Wind tunnel modelling of stack gas dispersion – Difficulties and aproximations", Wind Engineering, Proceedings of the fifth International Conference, Fort Collins, Colorado, USA, 1980, Ed. by J. E. Cermak, Pergamon Press Ltd.

[7] Y. Gagne, M. Marchand and B. Castaing, A new representation of energy spectra in fully developed turbulence, Applied Scientific Research, Volume 51, 99-103, 1993.

[8] Arya, S. P., "Atmospheric boundary layers over homogeneous terrain", Engineering Meteorology, Ed. by E. J. Plate, Elsevier Scientific Publishing Company, Amsterdam, 1982: 233-266.

[9] Cook, N.J., "A Boundary Layer Wind Tunnel for Building Aerodynamics", Journal of Industrial Aerodynamics, 1975, 1: 3-12.

[10] Sykes, D.M., "A New Wind Tunnel for Industrial Aerodynamics", Journal of Industrial Aerodynamics 1977, 2: 65-78.

[11] Greenway, M., Wood, C., "The Oxford University 4 m × 2 m Industrial Aerodynamics Wind Tunnel" Journal of Industrial Aerodynamics 1979, 4, 43-70.

[12] Blessmann, J., "The Boundary Layer TV-2 Wind Tunnel of the UFGRS", Journal of Wind Engineering and Industrial Aerodynamics, 1982, 10: 231-248.

[13] Hansen, S., Sorensen, E., "A New Boundary Layer Wind Tunnel at the Danish Maritime Institute", Journal of Wind Engineering and Industrial Aerodynamics 1985, 18: 213-224.

[14] Wittwer, A. R., Loredo-Souza, A. M., Schettini, E. B. C., Laboratory evaluation of the urban effects on the dispersion process using a small-scale model. In: 13th International Conference on Wind Engineering, Amsterdam. Proceedings of the ICWE13, 2011.

[15] Loredo Souza, A. M., "The behaviour of transmission lines under high winds", Thesis – Doctor of Philosophy, The University of Western Ontario, London, Ontario, 1996.

[16] Counihan, J., "An Improved Method of Simulating an Atmospheric Boundary Layer in a Wind Tunnel", Atmospheric Enviroment 1969, 3: 197-214.

[17] Standen, N.M., "A Spire Array for Generating Thick Turbulent Shear Layers for Natural Wind Simulation in Wind Tunnels", National Research Council of Canada, NAE, Report LTR-LA-94, 1972.

[18] A. R. Wittwer, M. E. De Bortoli, M. B. Natalini, "The importance of velocity fluctuations analysis at the atmospheric boundary layer simulation in a wind tunnel", 2nd East European Conference on Wind Engineering, Proceedings, Vol. 2, pp. 385, Academy of Sciences of the Czech Republic, Institute of Theoretical and Applied Mechanics, Czech Republic, Prague, 1998.

[19] Wittwer A. R., Möller S. V., "Characteristics of the low speed wind tunnel of the UNNE", Journal of Wind Engineering & Industrial Aerodynamics, 2000, 84: 307-320.

[20] J. Marighetti, A. Wittwer, M. De Bortoli, B. Natalini, M. Paluch, M. B. Natalini, "Fluctuating and mean pressures measurements in a wind tunnel over a stadium covering", Journal of Wind Engineering & Industrial Aerodynamics, 2000, 84: 321-328.

[21] Centro de Investigación de los Reglamentos Nacionales de Seguridad para las Obras Civiles, Reglamento CIRSOC 102, "Acción del Viento sobre las Construcciones", IN-TI, Argentina, 1982.

[22] Liu, G., Xuan, J., Park, S., "A new method to calculate wind profile parameters of the wind tunnel boundary layer", Journal of Wind Engineering and Industrial Aerodynamics 2003, 91: 1155-1162.

[23] Robins, A., Castro, I., Hayden, P., Steggel, N., Contini, D., Heist, D., "A wind tunnel study of dense gas dispersion in a neutral boundary layer over a rough surface", Atmospheric environment 2001, 35: 2243-2252.

[24] Cook, N. J., "The designer's guide to wind loading of building structures", BRE, Building Research Establishment, London, UK, 1990.

[25] Press, W.H., Flannery, B.P., Teukolsky, S.A., Vetterling, W.T., "Numerical Recipes: The Art of Scientific Computing", Cambridge University Press, New York, 1990.

[26] Acevedo, O.C., Moraes, O.L.L., Degrazia, G.A., Medeiros, L.E., "Intermittency and the exchange of scalars in the nocturnal surface layer", Boundary-layer meteorology, 119: 41-55, 2006.

[27] Wittwer A. R., Welter G. S., Degrazia G. A., "Características espectrales de la turbulencia en vientos de capa superficial", Proceedings de 1er. Congreso Latinoamericano de Ingeniería del Viento, Montevideo, Uruguay, 4 - 6 nov. 2008.

[28] Wittwer A. R., Welter G. S., Degrazia G. A., Loredo-Souza A. M., Acevedo O. C., Schettini E. B. C., Moraes O. L. L., "Espectros de turbulência medidos na camada atmosférica superficial e em um túnel de vento de camada limite.", Ciência & Natura, Revista do Centro de Ciências Naturais e Exatas, UFSM, Brasil, V. Esp. 2007: 137-141.

[29] Welter, G. S., Wittwer, A. R., Degrazia, G.A., Acevedo, O.C., Moraes, O.L.L., Anfossi, D., "Measurements of the Kolmogorov constant from laboratory and geophysical wind data", Physica A: Statistical Mechanics and its Applications, 388 : 3745-3751, 2009.

[30] Degrazia G. A., Welter G. S., Wittwer A. R., Carvalho J., Roberti D. R., Acevedo O. C., Moraes O.L.L., Velho H.F.C., "Estimation of the Lagrangian Kolmogorov constant from Eulerian measurements for distinct Reynolds number with application to pollution dispersion model", Atmospheric Environment 42, pp. 2415–2423, 2008.

[31] Degrazia, G., Anfossi, D., Carvalho, J., Mangia, C., Tirabassi, T. & Campos Velho, H. Turbulence parameterisation for PBL dispersion models in all stability conditions, Atmospheric environment 34(21): 3575–3583, 2000.

[32] Kolmogorov, A. N. (1941a). Energy dissipation in locally isotropic turbulence, Dokl. Akad. Nauk SSSR, Vol. 32, pp. 19–21, 1941.

[33] Kolmogorov, A. N. (1941b). The Local Structure of Turbulence in Incompressible Viscous Fluid for Very Large Reynolds' Numbers, Dokl. Akad. Nauk SSSR, Vol. 30, pp. 301–305, 1941.

[34] Monin A., Yaglom A., Statistical fluid mechanics: Vol. 2. [S.l.]: MIT Press, 1975.

[35] Falkovich, G., Sreenivasan, K. R. Lessons from Hydrodynamic Turbulence, Phys. Today 59, 43 (2006), DOI:10.1063/1.2207037

[36] Frisch U. Turbulence: The Legacy of AN Kolmogorov. [S.l.]: Cambridge University Press, 1995.

Permissions

The contributors of this book come from diverse backgrounds, making this book a truly international effort. This book will bring forth new frontiers with its revolutionizing research information and detailed analysis of the nascent developments around the world.

We would like to thank N. A. Ahmed, for lending his expertise to make the book truly unique. He has played a crucial role in the development of this book. Without his invaluable contribution this book wouldn't have been possible. He has made vital efforts to compile up to date information on the varied aspects of this subject to make this book a valuable addition to the collection of many professionals and students.

This book was conceptualized with the vision of imparting up-to-date information and advanced data in this field. To ensure the same, a matchless editorial board was set up. Every individual on the board went through rigorous rounds of assessment to prove their worth. After which they invested a large part of their time researching and compiling the most relevant data for our readers. Conferences and sessions were held from time to time between the editorial board and the contributing authors to present the data in the most comprehensible form. The editorial team has worked tirelessly to provide valuable and valid information to help people across the globe.

Every chapter published in this book has been scrutinized by our experts. Their significance has been extensively debated. The topics covered herein carry significant findings which will fuel the growth of the discipline. They may even be implemented as practical applications or may be referred to as a beginning point for another development. Chapters in this book were first published by InTech; hereby published with permission under the Creative Commons Attribution License or equivalent.

The editorial board has been involved in producing this book since its inception. They have spent rigorous hours researching and exploring the diverse topics which have resulted in the successful publishing of this book. They have passed on their knowledge of decades through this book. To expedite this challenging task, the publisher supported the team at every step. A small team of assistant editors was also appointed to further simplify the editing procedure and attain best results for the readers.

Our editorial team has been hand-picked from every corner of the world. Their multi-ethnicity adds dynamic inputs to the discussions which result in innovative outcomes. These outcomes are then further discussed with the researchers and contributors who give their valuable feedback and opinion regarding the same. The feedback is then collaborated with the researches and they are edited in a comprehensive manner to aid the understanding of the subject.

Apart from the editorial board, the designing team has also invested a significant amount of their time in understanding the subject and creating the most relevant covers. They scrutinized every image to scout for the most suitable representation of the subject and create an appropriate cover for the book.

The publishing team has been involved in this book since its early stages. They were actively engaged in every process, be it collecting the data, connecting with the contributors or procuring relevant information. The team has been an ardent support to the editorial, designing and production team. Their endless efforts to recruit the best for this project, has resulted in the accomplishment of this book. They are a veteran in the field of academics and their pool of knowledge is as vast as their experience in printing. Their expertise and guidance has proved useful at every step. Their uncompromising quality standards have made this book an exceptional effort. Their encouragement from time to time has been an inspiration for everyone.

The publisher and the editorial board hope that this book will prove to be a valuable piece of knowledge for researchers, students, practitioners and scholars across the globe.

List of Contributors

Miguel A. González Hernández, Ana I. Moreno López, Artur A. Jarzabek and José M. Perales Perales
Polytechnic University of Madrid, Spain

Yuliang Wu and Sun Xiaoxiao
Beijing Institute of Technology, China

N. A. Ahmed
Aerospace Engineering, School of Mechanical and Manufacturing Engineering, University of New South Wales, Sydney, NSW, Australia

N. A. Ahmed
School Of Mechanical and Manufacturing Engineering, University of New South Wales, Sydney, NSW, Australia

R. Scott Van Pelt
United States Department of Agriculture – Agricultural Research Service (USDA-ARS), Big
Spring, Texas, USA

Ted M. Zobeck
USDA-ARS, Lubbock, Texas, USA

Josué Njock Libii
Indiana University-Purdue University Fort Wayne, Fort Wayne, Indiana, USA

Abdulaziz Almubarak
College of Technological Studies, Department of Chemical Engineering, Kuwait

Yuki Nagai
Sasaki Structural Consultants, Japan

Akira Okada, Naoya Miyasato, Masao Saitoh and Ryota Matsumoto
Nihon University, Japan

Yoshifumi Yokoi
National Defense Academy of Japan, Japan

Adrián Roberto Wittwer
Facultad de Ingeniería, Universidad Nacional del Nordeste, Argentina

Guilherme Sausen Welter
Laboratório Nacional de Computação Científica, Brasil

Acir M. Loredo-Souza
Universidade Federal de Rio Grande do Sul, Brasil

CPSIA information can be obtained
at www.ICGtesting.com
Printed in the USA
BVHW091611121118
532888BV00003B/27/P